Compact Textbooks in Mathematics

For further volumes:
http://www.springer.com/series/11225

Compact Textbooks in Mathematics

This textbook series presents concise introductions to current topics in mathematics and mainly addresses advanced undergraduates and master students. The concept is to offer small books covering subject matter equivalent to 2- or 3-hour lectures or seminars which are also suitable for self-study. The books provide students and teachers with new perspectives and novel approaches. They feature examples and exercises to illustrate key concepts and applications of the theoretical contents. The series also includes textbooks specifically speaking to the needs of students from other disciplines such as physics, computer science, engineering, life sciences, finance.

Belkacem Said-Houari

Differential Equations: Methods and Applications

 Springer

Belkacem Said-Houari
Mathematics and Natural Sciences Department
Alhosn University
Abu Dhabi
United Arab Emirates

ISSN 2296-4568
Compact Textbooks in Mathematics
ISBN 978-3-319-25734-1   ISBN 978-3-319-25735-8 (eBook)
DOI 10.1007/978-3-319-25735-8
Springer Cham Heidelberg New York Dordrecht London

Library of Congress Control Number: 2015958023

Mathematics Subject Classification (2010): 34-01, 34A30, 34A34

Cover design: deblik, Berlin

Printed on acid-free paper

Springer International Publishing is part of Springer Science+Business Media
(www.springer.com)

This work is dedicated to my parents Ahmed and Bakhta and my wife Nadia.

Preface

Nowadays, differential equations is the centerpiece in engineering, physics, mechanics, life sciences and in many areas of mathematical modeling.

Among the courses taken by graduate students in engineering and mathematics is *Differential Equations and Applications*. I have been teaching this course at Alhosn university in UAE since Spring 2014. So, some parts of this book have been taught to the engineering undergraduate students at Alhosn university.

In this book, I have attempted to give a brief and modern introduction to the subject of ordinary differential equations. I have presented first the roots of the mathematical methods used to solve different differential equations and then I have followed each section/chapter with exercises with detailed solutions. Most of the exercises can be found as theorems in other textbooks. In addition, I have given some exercises as a direct application to the main theorems. Any student with some knowledge of matrix algebra and calculus II, especially integrals and series can read this book easily.

This book contains material that traditionally found in the classical textbooks as well as recent developments. I have tried to present the course in a simple way by covering the mean ideas in differential equations. The idea of the book is to give students a simple and short text book that can be easily read and quickly reaches the main ideas. The main parts of the book aimed to show students how to construct solutions of different types of differential equations.

I have started in ► Chapter 1 by giving some models where we can find differential equations such as: physics, ecology, biology, geometry, mechanics and electricity. I have also presented the basic definitions of differential equations.

► Chapter 2 is devoted to the study of the different kinds of first order differential equations including: separable equations, linear equations, homogeneous equations, Bernoulli equations, exact equations and Riccati equations. For each type, we have given many examples and exercises with solutions in order to help students to learn how to write a typical solution for differential equations, since during my teaching experience and by supervising many PhD and Master students, I have seen that even some PhD students that find it hard to write an organized solution or to write good papers due to the lack of the methodology.

► Chapter 3 covers the linear second order differential equation with constant coefficient or variable coefficients. I have given the methods that allow us to deal with those differential equations.

► Chapter 4 is subjected to the study of the Laplace transform and its applications and its usefulness in solving initial value problems.

In ► Chapter 5, the power series method has been introduced, especially in order to tackle differential equations with variable coefficients.

► Chapter 6 is devoted to the solutions of systems of differential equations using matrix algebra. In this chapter, we presented a method that relies only on the knowledge of the eigenvalues of the matrix of the system in order to find a solution of the system of differential equations and we avoided the Jordan canonical form.

Finally, in ▶ Chapter 7, we studied the qualitative theory of differential equations, where we attempted to deduce the behavior of the solution without knowing the explicit solutions.

Two types of exercises are given in this book, those which illustrate the general theory and others designed to fill out the text material. I included detailed solutions to all the exercises.

This book is excellently suited for use as a textbook for an undergraduate class (of all disciplines) in ordinary differential equations.

Finally, in writing this book of this nature, no originality can be claimed. But great efforts have been made to present the subject as simple and clear as possible.

Abu Dhabi, February 20th, 2015   Belkacem Said-Houari

Contents

Modelling and Definitions

Belkacem Said-Houari

B. Said-Houari, *Differential Equations: Methods and Applications*, Compact Textbooks in Mathematics,
DOI 10.1007/978-3-319-25735-8_1, © Springer International Publishing Switzerland 2015

1.1   Background

In order to make a bridge between the mathematical theory and the different applications of differential equations in real life, we begin by giving simple applications of differential equations.

1.1.1   Free Fall

An object of mass m is released from a certain height h above the ground and falls under the force of gravity, see ◘ Fig. 1.1. We want to write a mathematical model that describe the change of the height h with respect to time t. We assume that at time $t = 0$, we know the initial height h_0 and the initial velocity of the object $v_0 = \frac{dh}{dt}|_{t=0}$. Newton's second law states that the mass of the object m times its acceleration $\frac{d^2h}{dt^2}$ equals the total forces acting on it.

Since we have a free fall and if we neglect the force of the air friction, we find:

$$m\frac{d^2h}{dt^2} = -mg, \tag{1.1}$$

where g is the gravity. Equation (1.1) is the mathematical model (*the differential equation*) describing the change of the height h with respect to the time t. If we divide both sides in (1.1) by m and integrate the result from 0 to t, we obtain

$$\int_0^t \frac{d^2h}{ds^2} ds = -\int_0^t g\,ds.$$

This gives

$$\frac{dh}{dt} - v_0 = -gt. \tag{1.2}$$

Integrating once again (1.2) from 0 to t, we obtain

$$\int_0^t \left(\frac{dh}{ds} - v_0\right) ds = -\int_0^t gs\,ds.$$

☐ **Fig. 1.1** An object of mass m falling from a height h above the ground

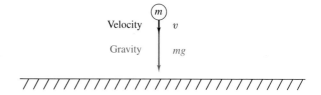

This leads to

$$h(t) = -\frac{gt^2}{2} + v_0 t + h(0). \tag{1.3}$$

Formula (1.3) is called the *solution* of the differential equation (1.1). It is obvious from (1.3) that the height $h(t)$ at any time t depends on the initial height h_0 and on the initial velocity v_0 but it does not depend on the mass m.

1.1.2 Population Dynamics and the Logistic Equation

Let $p(t)$ represent the total number of members of a population at time t (a population can be seen as any collection of objects we can count: animals, biological cells, automobiles, etc). If b represents the birth rate and d represents the death rate of the population $p(t)$, then the rate of change of $p(t)$ with respect to time is given by

$$\frac{dp(t)}{dt} = bp(t) - dp(t) = rp(t), \qquad r = b - d. \tag{1.4}$$

Equation (1.4) is a *differential equation*. Suppose that at time $t = 0$, the initial population is $p(0) = p_0$. Thus, equation (1.4) leads to

$$\frac{p'(t)}{p(t)} = r. \tag{1.5}$$

Integrating (1.5) from 0 to t, we get

$$\int_0^t \frac{p'(s)}{p(s)} ds = \int_0^t r \, ds.$$

This gives, since $p(t) > 0$,

$$\ln p(t) - \ln p_0 = rt.$$

Using the properties of the logarithm and taking the exponential of both sides in the above formula, we obtain

$$p(t) = p_0 e^{rt}. \tag{1.6}$$

Fig. 1.2 Population growth: $r > 0$

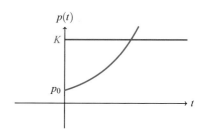

Fig. 1.3 The population growth in the logistic equation: $r > 0$

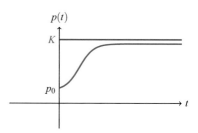

Consequently, the function $p(t)$ given by (1.6) is the *solution* of the differential equation (1.4). In equation (1.6) and if $r > 0$, meaning that the birth rate is greater than the death rate, then the population is growing exponentially fast. Conversely, if $r < 0$, meaning that the death rate is greater than the birth rate, then the population is decaying exponentially fast. If $r = 0$, then the population is constants.

It is clear from (1.6) that if $r > 0$, then the population $p(t)$ grows exponentially. But let us assume that the environment does support only a finite number K of the population, due to the lack of resources such as food, water, ... (see **Fig. 1.2**). Then, in this case the model (1.4) is not the right one. So, we want to look for an alternative equation that models the following situation: if $p(t)$ is much smaller than K, then we should have a growth rate as in (1.6), but once the population reaches the number K, then the growth rate should approach zero. See **Fig. 1.3**. Then, the model should be

$$\frac{dp(t)}{dt} = rp(t)\left(1 - \frac{p(t)}{K}\right). \tag{1.7}$$

Equation (1.7) is called the *logistic equation* and the number K is called the *carrying capacity* (the maximum sustainable population).

Equation (1.7) is a first order equation that can be solved in a similar manner to (1.4). Indeed, we have from (1.7)

$$\int \frac{dp}{p\left(1 - \frac{p}{K}\right)} = \int r\,dt.$$

We can easily see that

$$\frac{1}{p\left(1 - \frac{p}{K}\right)} = \frac{1}{p} + \frac{1/K}{1 - p/K}.$$

Taking the above identity into account, we obtain

$$\int \left(\frac{1}{p} + \frac{1/K}{1 - p/K}\right) dp = \int r\, dt. \tag{1.8}$$

Integrating both sides in (1.8), we obtain

$$\ln p - \ln(1 - p/K) = rt + A,$$

where A is the constant of integration. This leads to

$$\ln \frac{p}{1 - p/K} = rt + A.$$

If at $t = 0$, $p = p_0$, then we have

$$A = \ln\left(p_0/(1 - p_0/K)\right)$$

and therefore, we have from above

$$\frac{p}{1 - p/K} = \frac{p_0 e^r t}{1 - p_0/K}.$$

Solving the above equation for p, we get

$$p(t) = \frac{K}{1 + \left(\frac{K}{p_0} - 1\right)e^{-rt}}. \tag{1.9}$$

Now, it is clear that for t large enough, then from (1.9), we deduce that the population $p(t)$ goes to the maximum sustainable population K. If $p(t) < K$, then the function $p(t)$ grows exponentially for small t and as $p(t)$ become larger, then the growth diminishes as in ◻ Fig. 1.3.

1.1.3 The Lotka–Volterra Predator–Prey Model

Let N be the number of preys (rabbits for example) and P be the number of the predator (foxes for example). Let a be the growth rate of the prey, b be the negative effect of the predator on the prey, c be the benefit the prey gives to the predator and d be the decay rate of the predator. The mathematical model describing this situation is the system of two ordinary differential equations:

$$\begin{cases} \frac{dN}{dt} = N(a - bP), \\ \frac{dP}{dt} = P(cN - d). \end{cases} \tag{1.10}$$

System (1.10) is known as the *Lotka–Volterra model* for predator prey interaction. From system (1.10), we can see that in the absence of the predator ($P = 0$), the first equation leads to

$$\frac{dN}{dt} = Na. \tag{1.11}$$

The solution of (1.11) is

$$N = N_0 e^{at},$$

where N_0 is the number of the preys at time $t = 0$. This means that the number of the prey grows unbounded in an exponential manner in the absence of the predator.

Now, for $c = 0$, we see that the second equation in (1.10) gives:

$$\frac{dP}{dt} = -dP. \tag{1.12}$$

The solution of (1.12) gives

$$P(t) = P_0 e^{-dt},$$

where P_0 is the number of the predators at $t = 0$. This last equation shows that the number of predators decreases exponentially without the benefit they get from the preys. For more analysis on this model, see Exercise 7.2.

1.1.4 Geometry

Let C be a curve in the xy-plane. Let $A = (x, y)$ be a point on this curve. We want to find the set of all points A from the curve C such that the tangent line L to the curve C at A is perpendicular to the line connecting this point to the origin $O = (0, 0)$ (see ◘ Fig. 1.4).

The equation of the line L is

$$y = \frac{dy}{dx} x + B, \tag{1.13}$$

◘ **Fig. 1.4** The arc C

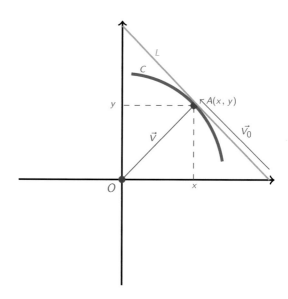

where B is a constants for any fixed point A. The components of the vector $\vec{V_0}$ parallel to L are $\vec{V_0} = \langle 1, \frac{dy}{dx} \rangle$. Also, the components of the vector $\overrightarrow{OA}$ are $\overrightarrow{OA} = \langle x, y \rangle$. Now, the vector $\overrightarrow{OA}$ is perpendicular to L if and only if it is perpendicular to $\vec{V_0}$, this means that the dot product between $\overrightarrow{OA}$ and $\vec{V_0}$ is zero. That is:

$$x \cdot 1 + y \cdot \frac{dy}{dx} = 0, \tag{1.14}$$

which can be rewritten as

$$\frac{dy}{dx} = -\frac{x}{y}. \tag{1.15}$$

Thus the set of points A described above should satisfy the *differential equation* (1.15). To solve (1.15), we rewrite it as

$$y\,dy = -x\,dx \tag{1.16}$$

and integrate both sides in (1.16), we find

$$y^2 + x^2 = c, \tag{1.17}$$

where c is a positive constant. Thus the curve C satisfying the above property is a circle of center $O = (0,0)$ and radius $\sqrt{c}$.

1.1.5 The Mass–Spring Oscillator

A damped mass–spring oscillator consists of a mass m attached to a nearby wall by means of a spring fixed at one end and it is free on the second end (Fig. 1.5). Our goal is to write a mathematical model that governs the motion of this oscillator, taking into account the forces acting on it due to the spring elasticity, the friction damping and the external influences. When the mass m is displaced from the equilibrium point O, the spring is compressed ($\ell < \ell_0$) or

 Fig. 1.5 Damped mass–spring oscillator

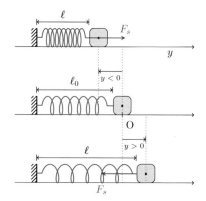

stretched ($\ell > \ell_0$) and it exerts a force that resist the displacement. This force is given by Hooke's law as

$$F_s = -ky,$$

where k is the spring stiffness and the negative sign represent the direction of the force with respect to the displacement y.

If $b \geq 0$ is the damping friction, the friction force is proportional to the velocity:

$$F_f = -b\frac{dy}{dt}.$$

We denote all the other forces which are external to the system by F_e.

Thus, Newton's second law says that the sum of all the forces acting on the system equals to the mass times the acceleration. This means

$$m\frac{dy^2}{dt^2} = F_s + F_f + F_e$$
$$= -ky - b\frac{dy}{dt} + F_e. \tag{1.18}$$

Consequently, the mathematical model describing the above system is the second order differential equation (1.18).

1.1.6 Electric Circuits

An electric circuit is a closed path in which electrons from a voltage or a current source flow. There are many types of electric circuits. Here we discuss two types: the RC and RLC circuits.

The RC Circuit An RC circuit is an electric circuit composed of a resistor and a capacitor driven by a voltage or a current source (for example a battery or a generator). Kirchhoff's law says: *The algebraic sum of all voltage drops around a closed circuit is zero.* The current source drives an electric charge Q and produce a current I. The current I and the charge Q are related by

$$I = \frac{dQ}{dt}. \tag{1.19}$$

The resistor opposes the current and dissipate the energy in the form of heat. The drop in voltage produced by the resistor of resistance R is given by Ohm's law as:

$$E_R = RI.$$

The capacitor stores charges and the associated electrostatic potential energy, and so it resists the flow of charges. A capacitor of capacitance C produce a voltage drop E_C given by:

$$E_C = \frac{Q}{C}.$$

If we let $E(t)$ denote the voltage supplied to the circuit at time t, then applying Kirchhoff's law to the RC circuit (■ Fig. 1.6), we get

$$E_R + E_C - E(t) = 0. \tag{1.20}$$

☐ **Fig. 1.6** RC circuit

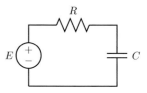

Using the above formulas, we may rewrite (1.20) as

$$R\frac{dQ}{dt} + \frac{Q}{C} = E(t). \tag{1.21}$$

Equation (1.21) is a first order differential equation which describe the change of the charge Q with respect to the time t.

The RLC Circuit The RLC circuit has a resistor, an inductor and a capacitor in series (or parallel) with voltage or a current source as in ☐ Fig. 1.7.

An inductor of inductance L produces a drop in voltage given by

$$E_L = L\frac{dI}{dt}.$$

Applying Kirchhoff's law to the RLC circuit, we have

$$E_L + E_R + E_C = E(t).$$

That is

$$L\frac{dI}{dt} + RI + \frac{Q}{C} = E(t) \tag{1.22}$$

If we assume that the voltage E supplied to the circuit is constant, then we may differentiate (1.22) with respect to t to get

$$L\frac{d^2I}{dt^2} + R\frac{dI}{dt} + \frac{1}{C}\frac{dQ}{dt} = 0. \tag{1.23}$$

Keeping in mind (1.19), then (1.23) takes the form

$$L\frac{d^2I}{dt^2} + R\frac{dI}{dt} + \frac{I}{C} = 0. \tag{1.24}$$

Equation (1.24) is a second order differential equation.

☐ **Fig. 1.7** RLC circuit

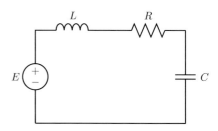

We point out that the RLC circuit has a remarkable similarity to the mass–spring oscilla-tor. Indeed, if we assume that $F_e = 0$ in (1.18), then equation (1.24) is precisely like (1.18) with I playing the role of y, L appearing in the place of the mass m, R has the role of the damping coefficient b and $1/C$ replacing the spring constant k.

1.2 Solutions and Initial Value Problems

Any differential equation of order n can be written in the following form:

$$F\left(x, y, \frac{dy}{dx}, \ldots, \frac{d^n y}{dx^n}\right) = 0, \tag{1.25}$$

where x is called the *independent* variable, y is called the *dependent* variable and F is a function that depends on x, y and the derivatives of y up to the order n. We assume that equation (1.25) holds for all x in an open interval $I = (a, b)$ where a could be $-\infty$ and b could be ∞.

Definition 1.2.1

The order of the differential equation (sometimes called degree) is the order of the highest derivative in the equation.

Example 1.1
The equation

$$\left(\frac{dy}{dx}\right)^2 - y + 2x = 0 \tag{1.26}$$

is an equation of first order, since the only derivative in (1.26) is dy/dx and the equation

$$\frac{d^2 y}{dx^2} - 4\frac{dy}{dx} + y - 5x^2 = 0 \tag{1.27}$$

is an equation of second order.

Definition 1.2.2 (Explicit solution)

A function $y = \phi(x)$ is an *explicit* solution to (1.25), if it satisfies the equation for all x in I. If the solution is not explicit, we say it is an *implicit* solution.

Example 1.2
For any constant $c > 0$, the function $\phi(x) = ce^x$ is an explicit solution to the differential equation

$$\frac{dy}{dx} = y \tag{1.28}$$

since it satisfies

$$\frac{d\phi(x)}{dx} = \phi(x),$$

for all x in $\mathbb{R}$.

Example 1.3

The solution of the differential equations (1.15) given by (1.17) is not an explicit solution. It is an implicit solution, since it is written in the form $\psi(x, y) = x^2 + y^2 - c = 0$, rather than $y = \phi(x)$.

Example 1.4

We consider the differential equation

$$\frac{d^2 y}{dx^2} + 3\frac{dy}{dx} + 2y = 0. \tag{1.29}$$

Show that for any two constants c_1 and c_2, the function

$$\phi(x) = c_1 e^{-x} + c_2 e^{-2x} \tag{1.30}$$

is an explicit solution to (1.29).

Solution

For $\phi(x)$ to be an explicit solution to (1.29), it has to satisfy the equation (1.29). That is

$$\frac{d^2\phi}{dx^2} + 3\frac{d\phi}{dx} + 2\phi = 0. \tag{1.31}$$

We have from above that

$$\frac{d\phi}{dx} = -c_1 e^{-x} - 2c_2 e^{-2x}$$

and thus

$$\frac{d^2\phi}{dx^2} = c_1 e^{-x} + 4c_2 e^{-2x}.$$

Plugging the above identities into (1.31), we deduce that (1.31) is satisfied for any $x \in \mathbb{R}$ and for any positive constants c_1 and c_2.

Definition 1.2.3 (Initial value problem)

An initial value problem for an nth order differential equation is the system containing equation (1.25) together with n equations as follows:

$$\begin{cases} F\left(x, y, \frac{dy}{dx}, \dots, \frac{d^n y}{dx^n}\right) = 0, \\ y(x_0) = y_0, \\ \frac{dy}{dx}(x_0) = y_1, \\ \vdots \\ \frac{d^{n-1} y}{dx^{n-1}}(x_0) = y_{n-1}, \end{cases} \tag{1.32}$$

where x_0 is a fixed point in some interval I and $y_0, \dots, y_{n-1}$ are given constants.

By solving the initial value problem (1.32), we mean find a solution to the first equation in (1.32) which satisfy the remaining other n equations in (1.32).

ⓘ **Remark 1.2.1** Usually, in an initial value problem, the number of the given constants $y_0, \ldots, y_{n-1}$ is equal to the order of the differential equation.

Example 1.5

Find the values of c_1 and c_2 such that the function $\phi(x)$ defined in (1.30) satisfies the initial value problem

$$
\begin{cases}
\dfrac{d^2y}{dx^2} + 3\dfrac{dy}{dx} + 2y = 0, \\[2mm]
y(0) = 1, \\[2mm]
\dfrac{dy}{dx}(0) = 2.
\end{cases}
\tag{1.33}
$$

Solution

We have already proved in Example 1.4, that $\phi(x)$ satisfies the first equation in (1.33) for all constants c_1, c_2 and for all x in $\mathbb{R}$. Now, we need to find c_1 and c_2 such that $\phi(x)$ also satisfies the initial conditions in (1.33). Indeed, $\phi(0) = 1$ gives

$$
c_1 + c_2 = 1
\tag{1.34}
$$

and $\frac{d\phi}{dx}(0) = 2$, leads to

$$
-c_1 - 2c_2 = 2
\tag{1.35}
$$

By solving the system of algebraic equations (1.34)–(1.35), we find $c_1 = 4$ and $c_2 = -3$. Thus,

$$
\phi(x) = 4e^{-x} - 3e^{-2x}
\tag{1.36}
$$

is the solution to the initial value problem (1.33).

Example 1.6

It is obvious that the function $\phi(x) = e^x$ is the solution to the initial value problem

$$
\begin{cases}
\dfrac{dy}{dx} - y = 0, \\[2mm]
y(0) = 1.
\end{cases}
\tag{1.37}
$$

1.2.1 First Order Initial Value Problem

Now, let us consider the first order initial value problem

$$
\begin{cases}
\dfrac{dy}{dx} = f(x, y), \\[2mm]
y(x_0) = y_0.
\end{cases}
\tag{1.38}
$$

As we have seen in Example 1.5 and Example 1.6, we found only one solution to the initial value problems (1.33) and (1.37). The question that we want to ask now is: *under which restrictions on the function $f(x, y)$ the initial value problem* (1.38) *has a unique solution?* The answer to this question is given in the following theorem:

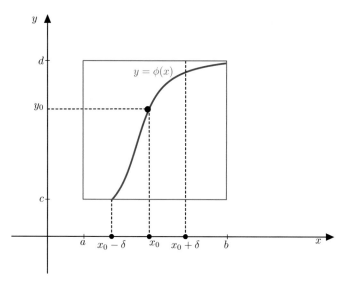

Illustration of Theorem 1.2.2. In the interval $(x_0 - \delta, x_0 + \delta)$, there is only one solution passing through the point (x_0, y_0)

Theorem 1.2.2 (Existence and uniqueness of solution)

Let us consider the initial value problem (1.38). Assume that the function f and its partial derivative with respect to y, $\partial f / \partial y$ are continuous functions in some rectangle

$$R = \{(x, y) : a < x < b, c < y < d\}.$$

Then for any (x_0, y_0) in R, there exists some $\delta > 0$, such that the initial value problem (1.38) has a unique solution $\phi(x)$ defined in the interval $x_0 - \delta < x < x_0 + \delta$, see ■ Fig. 1.8.

The proof of Theorem 1.2.2 can be found for instance in [15]. We omit its details.

1.3 Linear Differential Equations

In this section, we talk about a special class of differential equations, called *linear differential equations*. These are the equations in which the derivative of the highest order is a linear function of the lower order derivatives.

Definition 1.3.1 (Linear equation)

A linear differential equation of order n, is an equation of the form

$$a_n(x)\frac{d^n y}{dx^n} + a_{n-1}\frac{d^{n-1} y}{dx^{n-1}} + \ldots + a_1\frac{dy}{dx} + a_0(x)y = f(x), \tag{1.39}$$

where $a_n(x)$, $a_{n-1}(x), \ldots, a_1(x)$, $a_0(x)$ and $f(x)$ are functions depending on x only.

Example 1.7

The equation

$$x\frac{dy}{dx} + x^2 y = \frac{1}{x} \tag{1.40}$$

is a linear differential equation of first order, since it can be written in the form (1.39) with $n = 1$, $a_1(x) = x$, $a_0(x) = x^2$ and $f(x) = 1/x$.

Example 1.8

The equation

$$\frac{d^2 y}{dx^2} + \frac{dy}{dx} - y = 0, \tag{1.41}$$

is a linear differential equation of second order of the form (1.39) with $n = 2$, $a_2(x) = a_1(x) = 1$, $a_0(x) = -1$ and $f(x) = 0$.

Example 1.9

The equation

$$\frac{dy}{dx} + y^2 = 1, \tag{1.42}$$

is nonlinear, since it cannot be written in the form (1.39) due to the term y^2.

First-Order Differential Equations

Belkacem Said-Houari

B. Said-Houari, *Differential Equations: Methods and Applications*, Compact Textbooks in Mathematics,
DOI 10.1007/978-3-319-25735-8_2, © Springer International Publishing Switzerland 2015

This chapter is devoted to the study of first order differential equations. We first classify the type of the differential equation that we want to solve, then for each type we apply the appropriate method.

2.1 Separable Equations

One of the simplest differential equations that we can easily solve, by using integration, is the *separable* differential equation.

Definition 2.1.1 (Separable equation)

The first order equation

$$\frac{dy}{dx} = f(x, y) \tag{2.1}$$

is *separable* if it can be written in the following form:

$$h(y)dy = g(x)dx, \tag{2.2}$$

where the function $h(y)$ depends only on y and the function $g(x)$ depends only on x.

Example 2.1
The differential equation

$$\frac{dy}{dx} = (1 - x)y^2$$

is separable since it can be written in the form

$$\frac{dy}{y^2} = (1 - x)dx,$$

which is in the form (2.2) with $h(y) = 1/y^2$ and $g(x) = 1 - x$.

Example 2.2

The differential equation

$$\frac{dy}{dx} = 1 + x^2 y$$

is not separable, since it cannot be written in the form (2.2).

To solve a separable differential equation, we do the following:

1. Write the equation in the form

$$h(y)dy = g(x)dx. \tag{2.3}$$

2. Integrate both sides in (2.3), to obtain

$$\int h(y)dy = \int g(x)dx.$$

This gives

$$H(y) + c_1 = G(x) + c_2,$$

which leads to

$$H(y) = G(x) + C, \qquad C = c_2 - c_1. \tag{2.4}$$

The equation (2.4) gives an implicit solution to the separable equation (2.2).

3. Write the solution explicitly when possible.

Example 2.3

Solve the differential equation

$$(1 + x^2)^2 \frac{dy}{dx} + 2x + 2xy^2 = 0. \tag{2.5}$$

Solution

Equation (2.5) can be easily written in the form

$$\frac{1}{1+y^2}dy = \frac{-2x}{(1+x^2)^2}dx.$$

This means that equation (2.5) is separable. Thus, integrating both sides in the above formula, we get

$$\int \frac{1}{1+y^2}dy = \int \frac{-2x}{(1+x^2)^2}dx.$$

This gives:

$$\tan^{-1} y + c_1 = \frac{1}{x^2+1} + c_2. \tag{2.6}$$

Applying the tan function to both sides in (2.6), we obtain:

$$y(x) = \tan\left[\frac{1}{x^2+1} + C\right], \qquad C = c_2 - c_1. \tag{2.7}$$

Formula (2.7) is the explicit solution of (2.5).

Example 2.4

Solve the equation

$$x\frac{dy}{dx} - (x+1)y = 0 \tag{2.8}$$

in the interval $I = \mathbb{R} - \{0\}$.

Solution

Equation (2.8) is equivalent to

$$\frac{1}{y}dy = \frac{x+1}{x}dx. \tag{2.9}$$

Therefore equation (2.8) is separable. Integrating both sides in (2.9), we get

$$\int \frac{1}{y}dy = \int \left(1 + \frac{1}{x}\right)dx.$$

This leads to, after performing the integration,

$$\ln|y| = x + \ln|x| + c_1.$$

Taking the exponential of both sides in the last identity, we obtain

$$|y| = e^{c_1}xe^x.$$

This leads to the solution of (2.8):

$$y(x) = Cxe^x, \qquad C = \pm e^{c_1}.$$

Example 2.5

Solve the differential equation

$$x\frac{dy}{dx} - y = 0 \tag{2.10}$$

in the interval $I = \mathbb{R} - \{0\}$.

Solution

Equation (2.10) is a separable equation since we can write it as

$$\frac{1}{y}dy = \frac{1}{x}dx.$$

Integrating both sides in the above equation, we find

$$\int \frac{1}{y}dy = \int \frac{1}{x}dx.$$

This yields

$$\ln|y| + c_1 = \ln|x| + c_2. \tag{2.11}$$

Taking the exponential of both sides in (2.11), we find

$$y = Cx, \qquad C = \pm e^c, \qquad c = c_2 - c_1.$$

Example 2.6

Find the solution to the initial value problem

$$\begin{cases} \dfrac{dy}{dx} = \left(1 - \dfrac{x}{2}\right) y^2, \\ y(0) = 1. \end{cases} \tag{2.12}$$

Solution

We write the first equation in (2.12) as

$$\frac{dy}{y^2} = \left(1 - \frac{x}{2}\right) dx. \tag{2.13}$$

Integrating both sides in (2.13), we find

$$-\frac{1}{y} = x - \frac{x^2}{4} + C.$$

In order to find the constant C in the above formula, we use the initial condition in (2.12), to get $C = -1$. This gives the unique solution

$$y(x) = \frac{4}{(x-2)^2}$$

to the initial value problem (2.12).

2.1.1 Exercises

Exercise 2.1

Solve the following differential equation

$$\frac{dy}{dx} \sin x = y, \tag{2.14}$$

in the interval $I = (0, \pi)$.

Solution

We write equation (2.14) in the form

$$\frac{dy}{y} = \frac{dx}{\sin x}. \tag{2.15}$$

We integrate both sides in (2.15), we obtain:

$$\int \frac{dy}{y} = \int \frac{dx}{\sin x}. \tag{2.16}$$

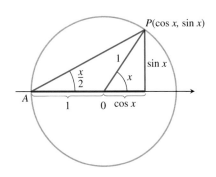

To compute the integral in the right-hand side of (2.16), we introduce the change of variable:

$$u = \tan \frac{x}{2}.$$

This gives,

$$dx = \frac{2du}{1+u^2}. \tag{2.17}$$

On the other hand, we have

$$\tan(x/2) = \frac{\sin x}{1 + \cos x}.$$

By using the fact that

$$\cos x = 2\cos^2(x/2) - 1 = \frac{1-u^2}{1+u^2},$$

we obtain, from above

$$\sin x = \frac{2u}{1+u^2}. \tag{2.18}$$

Plugging (2.17) and (2.18) into (2.16), we obtain

$$\int \frac{dy}{y} = \int \frac{du}{u}.$$

This gives

$$\ln|y| = \ln|u| + c.$$

Taking the exponential of both sides in the above equation, we get

$$y = Cu, \qquad C = \pm e^c.$$

Replacing u with $\tan(x/2)$, we get

$$y(x) = C \tan \frac{x}{2},$$

as the solution of (2.14).

Exercise 2.2

Solve the differential equation

$$x^2 y \frac{dy}{dx} = \sqrt{1-y^2}. \tag{2.19}$$

Solution
First, we have the trivial solutions $y(x) = \pm 1$.

Second, if y is not the constant solution, then, we may write equation (2.19) as

$$\frac{y}{\sqrt{1-y^2}} dy = \frac{dx}{x^2}.$$

Thus, by integrating both sides, we obtain

$$-\sqrt{1-y^2} = -\frac{1}{x} + C. \tag{2.20}$$

The formula (2.20) represents an implicit solution to the separable differential equation (2.19).

Exercise 2.3

Find the solutions to the differential equation

$$(x^2 - 4)y' = y \tag{2.21}$$

in the interval $I = \mathbb{R} - \{-2, 2\}$.

Solution

As before, we can rewrite (2.21) as

$$\frac{dy}{y} = \frac{dx}{(x^2 - 4)}$$

$$= \left(\frac{1}{4(x-2)} - \frac{1}{4(x+2)} \right) dx. \tag{2.22}$$

Integrating both sides in (2.22), we find

$$\ln|y| = \frac{1}{4} \left(\ln|x-2| - \ln|x+2| \right) + c$$

$$= \ln \left| \frac{x-2}{x+2} \right|^{1/4} + c.$$

Applying the exponential function to both sides in the above formula, we obtain the solution of (2.21) to be

$$y(x) = C \left| \frac{x-2}{x+2} \right|^{1/4}, \qquad C = \pm e^c.$$

Exercise 2.4

Solve the differential equation

$$(1 + x^2)y' + 3xy = 0. \tag{2.23}$$

Solution

As before, the equation (2.23) is a separable differential equation and it can be written in the form

$$-\frac{1}{y} dy = \frac{3x}{1 + x^2} dx. \tag{2.24}$$

Integrating both sides in (2.24), we find

$$-\ln|y| = \frac{3}{2} \ln(1 + x^2) + c.$$

Using the logarithm properties and applying the exponential function to both sides, we obtain

$$y(x) = \frac{C}{(1 + x^2)^{3/2}}, \qquad C = \pm e^c.$$

Consequently, the above formula of y represents all the explicit solutions to the differential equation (2.23).

Exercise 2.5

Find all the explicit solutions to the differential equation

$$(1 + x^2)y' = 1 + y^2. \tag{2.25}$$

Solution

We rewrite equation (2.25) as

$$\frac{dy}{1 + y^2} = \frac{dx}{1 + x^2}.$$

Integrating both sides in the above equation, we find

$$\tan^{-1} y = \tan^{-1} x + C. \tag{2.26}$$

This is an implicit solution to the equation (2.25). In order to get y explicitly as a function of the independent variable x, we use the formula:

$$\tan(a - b) = \frac{\tan a - \tan b}{1 + \tan a \tan b},$$

to obtain, with $y = \tan a$ and $x = \tan b$,

$$\tan^{-1} y - \tan^{-1} x = \tan^{-1} \frac{y - x}{1 + xy} + k\pi, \qquad k \in \mathbb{Z}.$$

Plugging this last formula into (2.26), we obtain

$$\tan^{-1} \frac{y - x}{1 + xy} = C - k\pi,$$

which gives by taking the tangent of both sides,

$$\frac{y - x}{1 + xy} = D, \qquad D = \tan(C - k\pi).$$

Consequently, we obtain

$$y(x) = \frac{x + D}{1 - Dx}. \tag{2.27}$$

as the solution of (2.25).

2.2 Exact Equations

In this section, we study the *exact* differential equation, which is a first-order differential equation where its solution is in the form $f(x, y) = c$. Thus, suppose that we have a family of curves $f(x, y) = c$. Then, its differential equation can be written in the form

$$\frac{d}{dx} f(x, y) = 0.$$

2

This leads to

$$\frac{\partial f}{\partial x} + \frac{\partial f}{\partial y}\frac{dy}{dx} = 0. \tag{2.28}$$

Hence, the slope of the curve $f(x, y) = c$ is defined by

$$\frac{dy}{dx} = -\frac{\partial f/\partial x}{\partial f/\partial y}.$$

Multiplying (2.28) by dx, we get the total differential of f

$$df := \frac{\partial f}{\partial x}dx + \frac{\partial f}{\partial y}dy = 0. \tag{2.29}$$

Suppose that we reverse this logic and begin with the differential equation

$$M(x, y)dx + N(x, y)dy = 0. \tag{2.30}$$

If there happens to exist a function $f(x, y)$ such that

$$\frac{\partial f}{\partial x}(x, y) = M(x, y) \quad \text{and} \quad \frac{\partial f}{\partial y}(x, y) = N(x, y), \tag{2.31}$$

then, equation (2.30) can be written in the form (2.29) and its general solution is

$$f(x, y) = c.$$

Definition 2.2.1 (Exact equation)

The differential form

$$M(x, y)dx + N(x, y)dy$$

is said to be *exact* in a rectangle

$$R = \{(x, y) : a < x < b, c < y < d\},$$

if there is a function $f(x, y)$ such that

$$\frac{\partial f}{\partial x}(x, y) = M(x, y) \quad \text{and} \quad \frac{\partial f}{\partial y}(x, y) = N(x, y),$$

for all (x, y) in R.

If $M(x, y)dx + N(x, y)dy$ is an exact differential form, then the equation

$$M(x, y)dx + N(x, y)dy = 0,$$

is called an *exact equation.*

Example 2.7

The differential equation

$$y\,dx + x\,dy = 0$$

is an exact differential equation since its left-hand side is the total differential of the function $f(x, y) = xy$. Then, its general solution is of the form $xy = c$.

How to find a function $f(x, y)$ satisfying (2.31) is not always obvious. So, we need a test for exactness and a method for finding the function f. Thus, we have the following theorem.

Theorem 2.2.1

Assume that the first partial derivatives of $M(x, y)$ and $N(x, y)$ are continuous in the rectangle

$$R = \{(x, y) : a < x < b, \quad c < y < d\}.$$

Then, the equation

$$M(x, y)dx + N(x, y)dy = 0,$$

is an exact differential equation in R if and only if the condition

$$\frac{\partial M}{\partial y}(x, y) = \frac{\partial N}{\partial x}(x, y) \tag{2.32}$$

holds for all $(x, y) \in R$.

Proof

First, we need to prove that (2.32) is a necessary condition for exactness. Indeed, we know from elementary calculus that the mixed second partial derivatives of $f(x, y)$ are equal if they are continuous. That is

$$\frac{\partial^2 f}{\partial x \partial y}(x, y) = \frac{\partial^2 f}{\partial y \partial x}(x, y). \tag{2.33}$$

Keeping in mind (2.31), then the above equation leads to (2.32).

Second, we need to show that condition (2.31) is also a sufficient condition for exactness. That is we shall prove that (2.31) enables us to construct a function $f(x, y)$ that satisfies (2.31). Indeed, integrating the first equation in (2.31) with respect to x, we get

$$f(x, y) = \int M(x, y)dx + g(y). \tag{2.34}$$

Notice that instead of a constant of integration C, we put a function $g(y)$ since it must disappear under the derivative with respect to x. Now, we need to choose our function $g(y)$ such that the partial derivative of (2.34) with respect to y gives the second equation in (2.31). That is

$$\frac{\partial}{\partial y} \int M(x, y)dx + g'(y) = N(x, y).$$

Consequently,

$$g'(y) = N(x, y) - \frac{\partial}{\partial y} \int M(x, y)dx.$$

To find $g(y)$, we need just to integrate the above formula with respect to y to get

$$g(y) = \int \left(N(x, y) - \frac{\partial}{\partial y} \int M(x, y)dx \right) dy. \tag{2.35}$$

Now, we need to verify that the above integrand is a function of y only. This means that the derivative of the above integrand with respect to x is 0. We have

$$\frac{\partial}{\partial x} \left(N(x, y) - \frac{\partial}{\partial y} \int M(x, y)dx \right) = \frac{\partial N}{\partial x}(x, y) - \frac{\partial^2}{\partial x \partial y} \int M(x, y)dx$$

$$= \frac{\partial N}{\partial x}(x, y) - \frac{\partial^2}{\partial y \partial x} \int M(x, y)dx$$

$$= \frac{\partial N}{\partial x}(x, y) - \frac{\partial M}{\partial y}(x, y)$$

$$= 0,$$

according to (2.32). This finishes the proof of Theorem 2.2.1.

Summary of the method To solve a differential equation of the form

$$M(x, y)dx + N(x, y)dy = 0, \tag{2.36}$$

do the following:
1. Show that (2.36) is an exact equation by verifying the condition (2.32).
2. Put

$$\frac{\partial f}{\partial x}(x, y) = M(x, y)$$

and integrate the above equation with respect to x, to get (2.34).
3. Take the partial derivative of (2.34) with respect to y and put it equal to $N(x, y)$ to find $g(y)$ as in (2.35).
4. Substituting $g(y)$ into (2.34) to get $f(x, y)$.
5. Once we get $f(x, y)$, then the solution of the differential equation (2.36) is $f(x, y) = c$.

2.2.1 Exercises

Exercise 2.6

Find the general solution of the differential equation

$$y dx + \left(x + \frac{2}{y} \right) dy = 0. \tag{2.37}$$

Solution

Equation (2.37) is written in the form (2.36) with $M(x, y) = y$ and $N(x, y) = x + 2/y$. Since

$$\frac{\partial M}{\partial y}(x, y) = \frac{\partial N}{\partial x}(x, y) = 1,$$

then equation (2.37) is an exact equation. Thus, we may search for its general solution in the form $f(x, y) = c$, where $f(x, y)$ satisfies (2.31).

Since

$$\frac{\partial f}{\partial x}(x, y) = y,$$

then an integration of the above formula with respect to x gives

$$f(x, y) = xy + g(y). \tag{2.38}$$

Taking the derivative of this last equality with respect to y, we obtain

$$x + g'(y) = N(x, y) = x + \frac{2}{y}.$$

Consequently,

$$g'(y) = \frac{2}{y}.$$

Integrating the above equation with respect to y, we find

$$g(y) = \ln y^2.$$

(We omitted the constant of integration, since it does not affect the final result). Plugging $g(y)$ into (2.38), we obtain the solution of (2.37), written implicitly in the form

$$xy + \ln y^2 = c.$$

Exercise 2.7

Find the general solution of the differential equation

$$(y - x^3)dx + (x + y^3)dy = 0. \tag{2.39}$$

Solution

We can easily write equation (2.39) in the form (2.36) with $M(x, y) = y - x^3$ and $N(x, y) = x + y^3$. Since

$$\frac{\partial M}{\partial y}(x, y) = \frac{\partial N}{\partial x}(x, y) = 1,$$

then equation (2.36) is an exact equation.

We put

$$\frac{\partial f}{\partial x}(x, y) = y - x^3.$$

Then an integration of the above formula with respect to x gives

$$f(x, y) = xy - \frac{x^4}{4} + g(y). \tag{2.40}$$

Taking the derivative of (2.40) with respect to y, we obtain

$$x + g'(y) = N(x, y) = x + y^3.$$

Therefore,

$$g'(y) = y^3.$$

An integration of the above equation with respect to y gives

$$g(y) = \frac{y^4}{4}.$$

Consequently, by inserting $g(y)$ into (2.40), then we deduce that the solution of (2.39) is given by

$$4xy - x^4 + y^4 = c,$$

where c is a constant.

2.3   First Order Linear Equations – Method of Integrating Factors

In this section, we introduce a method of finding a solution to a linear differential equation of first order. As we have seen in Definition 1.3.1, the first order linear equation has the form

$$a_1(x)\frac{dy}{dx} + a_0(x)y = b(x), \tag{2.41}$$

where $a_1(x) \neq 0$, $a_0(x)$ and $b(x)$ are functions depending on the variable x only.

Before, investigating the general method of solving equation (2.41), let us first discuss the following two particular cases:

- **Case 1.** $a_0(x) = 0$. In this case, equation (2.41) takes the form

$$a_1(x)\frac{dy}{dx} = b(x), \tag{2.42}$$

which is a separable equation, since it can be rewritten in the form

$$dy = \frac{b(x)}{a_1(x)}dx. \tag{2.43}$$

Thus, using the method in ▶ Sect. 2.1 and integrating both sides in (2.43), we obtain

$$y(x) = \int \frac{b(x)}{a_1(x)}dx + C.$$

▬ **Case 2.** $a_0(x) = a_1'(x)$. In this case, equation (2.41) reads as

$$\underbrace{a_1(x)\frac{dy}{dx} + a_1'(x)y}_{= \frac{d}{dx}(a_1(x)y)} = b(x). \tag{2.44}$$

This gives

$$\frac{d}{dx}(a_1(x)y) = b(x). \tag{2.45}$$

Integrating both sides in (2.45) with respect to x, we obtain

$$\int \frac{d}{dx}(a_1(x)y)dx = \int b(x)dx.$$

Thus, we have

$$a_1(x)y = \int b(x)dx + C,$$

which gives

$$y(x) = \frac{1}{a_1(x)}\left(\int b(x)dx + C\right). \tag{2.46}$$

Now, to introduce the method of *integrating factor*, we write equation (2.41) as

$$\frac{dy}{dx} + p(x)y = q(x), \tag{2.47}$$

which can be obtained from (2.41) with $p(x) = a_0(x)/a_1(x)$ and $q(x) = b(x)/a_1(x)$. The idea of this method is to multiply both sides in (2.47) by a function $\mu(x)$ and choose our function $\mu(x)$ such that the obtained term in the left-hand side of the resulting equation equals to $\frac{d}{dx}(\mu(x)y)$. Indeed, we multiply (2.47) by $\mu(x)$ to get

$$\mu(x)\frac{dy}{dx} + \mu(x)p(x)y = \mu(x)q(x). \tag{2.48}$$

This implies

$$\frac{d}{dx}(\mu(x)y) - \mu'(x)y + \mu(x)p(x)y = \mu(x)q(x). \tag{2.49}$$

Therefore, our goal is to choose $\mu(x)$ such that

$$-\mu'(x)y + \mu(x)p(x)y = 0.$$

That is, we select $\mu(x)$ so that

$$\mu'(x) = p(x)\mu(x), \tag{2.50}$$

and therefore, we obtain from (2.49)

$$\frac{d}{dx}(\mu(x)y) = \mu(x)q(x),$$ (2.51)

which can be easily solved as in Case 2 above, to get

$$y(x) = \frac{1}{\mu(x)}\left(\int \mu(x)q(x)dx + C\right).$$ (2.52)

Now, to find the solution y in (2.52) it is enough to find one function $\mu(x)$ satisfying (2.50). This can be easily done by solving the differential equation (2.50) to get one of the solutions to be

$$\mu(x) = e^{\int p(x)dx}.$$

Summary of the method From above and in order to solve the first order linear differential equation of the form (2.41), do the following:
1. Write equation (2.41) in the form (2.47), (the $p - q$ from).
2. Find the function $\mu(x) = e^{\int p(x)dx}$.
3. Multiply (2.47) by $\mu(x)$ to get the equation (2.51).
4. Integrate equation (2.51) and find y as a function of x as in (2.52).

Example 2.8
Solve the linear differential equation

$$\frac{dy}{dx} + 2xy = x.$$ (2.53)

Solution
In equation (2.53) the independent variable is x and the equation is in the form (2.47) with $p(x) = 2x$ and $q(x) = x$.
    Next, we need to compute

$$\mu(x) = e^{\int p(x)dx} = e^{\int 2x dx}$$
$$= e^{x^2}.$$

Now, we multiply equation (2.53) by $\mu(x)$ to obtain

$$\frac{d}{dx}\left(e^{x^2}y\right) = xe^{x^2}.$$

Integrating both sides in the above formula with respect to x, we get

$$\int \frac{d}{dx}\left(e^{x^2}y\right)dx = \int xe^{x^2}dx.$$

This leads to

$$e^{x^2}y = \frac{1}{2}e^{x^2} + C.$$

Finally, we get

$$y(x) = \frac{1}{2} + Ce^{-x^2}.$$

Example 2.9

Solve the linear differential equation

$$\frac{dy}{dx} + y = x^2.$$ (2.54)

Solution

Equation (2.54) is written in the form (2.47) with $p(x) = 1$ and $q(x) = x^2$. We can easily find

$$\mu(x) = e^{\int dx} = e^x.$$

Multiplying equation (2.54) by e^x, we get

$$\frac{d}{dx}(e^x y) = x^2 e^x.$$ (2.55)

Integrating (2.55) with respect to x, we obtain

$$e^x y + c_1 = \int x^2 e^x dx.$$ (2.56)

Using the integration by parts twice to find

$$\int x^2 e^x dx = (x^2 - 2x + 2)e^x + c_2.$$

Inserting this last formula into (2.56), we find that the solution of (2.54) is

$$y(x) = x^2 - 2x + 2 + Ce^{-x}, \qquad C = c_2 - c_1.$$

2.3.1 Exercises

Exercise 2.8

Find the solutions of the differential equation

$$x\frac{dy}{dx} - 2y = \frac{1}{2}x^3$$ (2.57)

in the interval $I = \mathbb{R} - \{0\}$.

Solution

For all x in I, we can write equation (2.57) in the $p - q$ form (2.47) as

$$\frac{dy}{dx} - \frac{2}{x}y = \frac{1}{2}x^2,$$ (2.58)

with $p(x) = -2/x$ and $q(x) = \frac{1}{2}x^2$. Next, we put

$$\mu(x) = e^{\int p(x)dx} = e^{-2\ln|x|}$$

$$= \frac{1}{x^2}.$$

Now, we multiply equation (2.58) by $\mu(x) = 1/x^2$, we get

$$\frac{d}{dx}\left(\frac{y}{x^2}\right) = \frac{1}{2}.$$

Integrating both sides with respect to x in the above equation, we find

$$y(x) = \frac{x^3}{2} + Cx^2.$$

Exercise 2.9

Solve the linear differential equation

$$\frac{dy}{dx}\cos x - y \sin x = \sin 2x \tag{2.59}$$

in the interval $I = (0, \pi/2)$.

Solution

First, we write equation (2.59) in the $p - q$ form as

$$\frac{dy}{dx} - y \tan x = 2 \sin x, \tag{2.60}$$

where we used the fact that $\sin 2x = 2 \sin x \cos x$.

Now, we compute $\mu(x)$ as:

$$\mu(x) = e^{-\int \tan x \, dx} = e^{\ln |\cos x|}$$
$$= |\cos x|$$
$$= \cos x,$$

since $x \in (0, \pi/2)$. Now, we multiply equation (2.60) by $\cos x$, we obtain

$$\frac{d}{dx}(y \cos x) = \sin 2x.$$

Integrating both sides with respect to x in the above equation, we obtain

$$y(x) = \frac{1}{\cos x}\left(C - \frac{\cos 2x}{2}\right).$$

Exercise 2.10

Find the solutions of the linear differential equation

$$(1 - x^2)\frac{dy}{dx} + 2y = x + 2 \tag{2.61}$$

in the interval $I = \mathbb{R} - \{-1, 1\}$.

Solution

For all x in I, we rewrite equation (2.61) in the $p - q$ form as follows:

$$\frac{dy}{dx} + \frac{2}{1 - x^2}y = \frac{x + 2}{1 - x^2}. \tag{2.62}$$

Therefore, the function $\mu(x)$ is defined by

$$\mu(x) = e^{\int \frac{2}{1-x^2}dx}.$$

It is not hard to see that $\frac{2}{1-x^2} = \frac{1}{1+x} + \frac{1}{1-x}$. This gives

$$\int \frac{2}{1 - x^2}dx = \int \left(\frac{1}{1 + x} + \frac{1}{1 - x}\right)dx$$
$$= \ln|1 + x| - \ln|1 - x| = \ln\left|\frac{1 + x}{1 - x}\right|.$$

Thus,

$$\mu(x) = \left|\frac{1+x}{1-x}\right|.$$

We may choose $\mu(x) = \frac{1+x}{1-x}$, an multiply (2.62) by $\mu(x)$, we obtain

$$\frac{d}{dx}\left(\frac{1+x}{1-x}y\right) = \frac{x+2}{(1-x)^2}. \tag{2.63}$$

Now, the right-hand side in the above equation can be written as:

$$\frac{x+2}{(1-x)^2} = \frac{1}{x-1} + \frac{3}{(x-1)^2}.$$

Plugging this into (2.63) and integrating both sides with respect to x, we find

$$\frac{1+x}{1-x}y + c_1 = \int\left(\frac{1}{x-1} + \frac{3}{(x-1)^2}\right)dx$$

$$= \ln|x-1| + \frac{3}{1-x} + c_2.$$

This last formula yields

$$y(x) = \frac{3}{1+x} + \frac{1-x}{1+x}\left[\ln|x-1| + C\right], \quad C = c_2 - c_1.$$

Exercise 2.11
Find the solutions of the linear differential equation

$$\sqrt{x}\frac{dy}{dx} + y = 1 \tag{2.64}$$

in the interval $I = (0, +\infty)$.

Solution
For all x in I, we write equation (2.64) in the $p - q$ form as

$$\frac{dy}{dx} + \frac{1}{\sqrt{x}}y = \frac{1}{\sqrt{x}}. \tag{2.65}$$

Thus, we find $\mu(x)$ as

$$\mu(x) = e^{\int\frac{1}{\sqrt{x}}dx}$$
$$= e^{2\sqrt{x}}.$$

We multiply (2.65) by $\mu(x)$, we find

$$\frac{d}{dx}\left(e^{2\sqrt{x}}y\right) = \frac{e^{2\sqrt{x}}}{\sqrt{x}}.$$

This gives, by integrating both sides with respect to x in the above equation,

$$e^{2\sqrt{x}}y = e^{2\sqrt{x}} + C.$$

Consequently, the solution of (2.64) is given by

$$y(x) = 1 + Ce^{-2\sqrt{x}}.$$

Exercise 2.12

Solve in $I = (-\pi/2, \pi/2)$ the differential equation

$$\frac{dy}{dx} - y \tan x = \frac{1}{1 + \cos x}. \tag{2.66}$$

Solution

Equation (2.66) is already in the $p - q$ form. We compute $\mu(x)$ as in Exercise 2.9, to get

$$\mu(x) = e^{-\int \tan x\, dx} = |\cos x|.$$

We may choose $\mu(x) = \cos x$ (if we choose $\mu(x) = -\cos x$, we get the same result) and we multiply equation (2.66) by $\mu(x)$, we get

$$\frac{d}{dx}(y \cos x) = \frac{\cos x}{1 + \cos x}.$$

Integrating both sides in the above equation, we find

$$y \cos x + c_1 = \int \frac{\cos x}{1 + \cos x}\, dx. \tag{2.67}$$

To compute the integral in the right-hand side of (2.67), we use the same change of variable as in Exercise 2.1. Thus, we put $u = \tan(x/2)$, which gives

$$\cos x = \frac{1 - u^2}{1 + u^2} \quad \text{and} \quad dx = \frac{2du}{1 + u^2}.$$

Thus, we have

$$\begin{aligned}
\int \frac{\cos x}{1 + \cos x}\, dx &= \int \frac{1 - u^2}{1 + u^2}\, du \\
&= \int \left(\frac{2}{1 + u^2} - 1 \right) du \\
&= 2 \tan^{-1}(u) - u + c_2 \\
&= 2 \tan^{-1}(\tan(x/2)) - \tan(x/2) + c_2 \\
&= x - \tan(x/2) + c_2.
\end{aligned}$$

Inserting this last equality into (2.67), we obtain

$$y(x) = \frac{x - \tan(x/2) + C}{\cos x}, \qquad C = c_2 - c_1.$$

2.4   Substitutions and Transformations

In this section, we introduce some substitutions and transformations tools to solve some nonlinear first order differential equations. This includes the homogeneous equations and Bernoulli's equation.

2.4.1 Homogeneous Equations

Definition 2.4.1 (Homogeneous equation)

The nonlinear first order differential equation

$$\frac{dy}{dx} = f(x, y) \tag{2.68}$$

is *homogenous* if $f(x, y) = g(y/x)$. This means that the equation is homogeneous if the right-hand side in (2.68) can be expressed as a function of y/x.

Example 2.10

The differential equation

$$\frac{dy}{dx} = \frac{x + y}{x} \tag{2.69}$$

is homogeneous since the function

$$f(x, y) = \frac{x + y}{x}$$
$$= 1 + \frac{y}{x} = g(y/x),$$

where $g(v) = 1 + v$.

Example 2.11

The equation

$$(x^2 + y^2)dy + 2xydx = 0$$

is homogeneous since it can be written in the form

$$\frac{dy}{dx} = -\frac{2xy}{x^2 + y^2},$$

with

$$-\frac{2xy}{x^2 + y^2} = \frac{-2(y/x)}{1 + (y/x)^2}$$
$$= g(y/x),$$

where $g(v) = -\frac{2v}{1+v^2}$.

The method of solving a homogeneous first order differential equation In order to solve a homogeneous first order differential equation, we make the following substitution

$$v = \frac{y}{x}. \tag{2.70}$$

This gives

$$\frac{dy}{dx} = v + x\frac{dv}{dx}.$$

Consequently, equation (2.68) becomes

$$v + x\frac{dv}{dx} = g(v),$$

which can be rewritten as

$$\frac{dv}{g(v) - v} = \frac{dx}{x}. \tag{2.71}$$

Equation (2.71) is a separable equation with v being the dependent variable and x is the independent variable. We use the method described in ▶ Sect. 2.1 to solve it. After we find v from solving (2.71), then we need to express the solution of the original equation in terms of x and y.

Example 2.12
Find the solutions of the differential equation

$$\frac{dy}{dx} = \frac{x + y}{x}, \qquad x \neq 0. \tag{2.72}$$

Solution
As we have seen in Example 2.10, equation (2.72) is homogeneous. We put

$$v = \frac{y}{x}.$$

Then, we get

$$\frac{dy}{dx} = v + x\frac{dv}{dx}.$$

Inserting this last equality into (2.72), we obtain

$$v + x\frac{dv}{dx} = 1 + v.$$

This gives

$$dv = \frac{dx}{x}. \tag{2.73}$$

Equation (2.73) is a separable equation. Thus, integrating both sides in (2.73), we get

$$v = \ln|x| + C.$$

Replacing v with y/x, we obtain

$$y(x) = x \ln |x| + Cx.$$

This last formula represents the explicit solution of equation (2.72).

Example 2.13

Find the solution to the initial value problem

$$\begin{cases} (x^2 + y^2)dx - xy\,dy = 0, \\ y(1) = 2, \end{cases} \tag{2.74}$$

in the interval $I = \mathbb{R} - \{0\}$.

Solution

It is obvious that the first equation in (2.74) can be written as

$$\frac{dy}{dx} = \frac{x}{y} + \frac{y}{x} = g(y/x) \tag{2.75}$$

with $g(v) = v + \frac{1}{v}$. Consequently, equation (2.74) is homogeneous. Thus, we put

$$v = \frac{y}{x}.$$

Then, we find

$$\frac{dy}{dx} = v + x\frac{dv}{dx}.$$

Plugging this into equation (2.75), we obtain

$$v + x\frac{dv}{dx} = v + \frac{1}{v}.$$

This gives

$$v\,dv = \frac{dx}{x},$$

which is a separable equation. Integrating both sides in the above equation, we find

$$\frac{v^2}{2} = \ln |x| + c_1.$$

Consequently, keeping in mind that $v = y/x$, we obtain

$$y^2 = 2x^2 \ln |x| + Cx^2, \qquad C = 2c_1. \tag{2.76}$$

Using the initial value in (2.74), we obtain $C = 4$. Thus, the function

$$y^2(x) = 2x^2 \ln |x| + 4x^2$$

is the implicit solution to the initial value problem (2.74).

2.4.2 Bernoulli Equations

Definition 2.4.2

A *Bernoulli equation* is a first order nonlinear differential equation that can be written in the form

$$\frac{dy}{dx} + p(x)y = q(x)y^n, \tag{2.77}$$

where $p(x)$ and $q(x)$ are continuous functions on an interval $I = (a, b)$ and n is a real number.[1]

Example 2.14

The following differential equation

$$\frac{dy}{dx} - 5y = xy^3$$

is a Bernoulli differential equation with $n = 3$, $p(x) = -5$ and $q(x) = x$.

The method of solving Bernoulli equations The idea of solving the nonlinear first order Bernoulli equation is to transform it into a first order linear equation. Notice that if $n = 0$ or $n = 1$, then equation (2.77) is linear and we already know in ▶ Sect. 2.3 how to solve it. Therefore in this subsection, we are going to look for solution of (2.77) in the cases: $n \neq 0$ and $n \neq 1$.

Now, we may ask the question: *is it possible to transform the Bernoulli equation into a first order linear equation even if $n \neq 0$ and $n \neq 1$?* The answer to this question is yes, and in order to do this we have to remove y^n from the left-hand side of equation (2.77) by dividing each term in the equation by y^n. Thus, we get

$$y^{-n}\frac{dy}{dx} + p(x)y^{1-n} = q(x). \tag{2.78}$$

We may now introduce the substitution

$$v = y^{1-n}, \tag{2.79}$$

which gives, by using the chain rule,

$$\frac{dv}{dx} = (1-n)y^{-n}\frac{dy}{dx}. \tag{2.80}$$

Plugging (2.79) and (2.80) into (2.78), we find

$$\frac{1}{1-n}\frac{dv}{dx} + p(x)v = q(x). \tag{2.81}$$

---

[1] Note that if $n > 0$, then $y = 0$ is a trivial solution of (2.77). So, we exclude this solution in our discussion.

Equation (2.81) is a first order linear equation with the dependent variable v and the independent variable x. We apply the method of integrating factor described in Section ▶ Sect. 2.3 to solve it for v. After that we just write the solution using the variables x and y by exploiting the equation (2.79).

Example 2.15

Solve for x in $\mathbb{R} - \{0\}$ the differential equation

$$\frac{dy}{dx} - \frac{y}{x} = xy^2. \tag{2.82}$$

Solution

Equation (2.82) is a Bernoulli type equation with $n = 2$. Our goal is to transform it into a first order differential equation. Indeed, we put

$$v = \frac{1}{y},$$

then we get

$$\frac{dv}{dx} = -\frac{1}{y^2}\frac{dy}{dx}.$$

Now, dividing equation (2.82) by y^2, we obtain

$$\frac{1}{y^2}\frac{dy}{dx} - \frac{1}{xy} = x. \tag{2.83}$$

Using the above identities, then equation (2.83) takes the form

$$-\frac{dv}{dx} - \frac{1}{x}v = x. \tag{2.84}$$

Or equivalently,

$$\frac{dv}{dx} + \frac{1}{x}v = -x. \tag{2.85}$$

To solve equation (2.85) we use the method of integrating factor. The equation is already in the $p - q$ form with $p(x) = 1/x$ and $q(x) = -x$. We compute

$$\mu(x) = e^{\int \frac{1}{x}dx} = |x|.$$

Next, we choose $\mu(x) = x$ and multiply equation (2.85) by $\mu(x)$, we get

$$\frac{d}{dx}(xv) = -x^2.$$

Integrating both sides in the above equation with respect to x, we obtain

$$xv = -\frac{x^3}{3} + C.$$

Consequently,

$$v = -\frac{x^2}{3} + \frac{C}{x}. \tag{2.86}$$

Keeping in mind that $v = 1/y$, then the solution of (2.15) is

$$y(x) = \frac{1}{-\frac{x^2}{3} + \frac{C}{x}}.$$

Example 2.16

Solve for $x \neq 0$ the following differential equation

$$x\frac{dy}{dx} + 6y = 3xy^{4/3}. \tag{2.87}$$

Solution

Equation (2.87) is a Bernoulli type equation with $n = 4/3$. We divide each term by $y^{4/3}$, we obtain

$$x\frac{dy}{dx}y^{-4/3} + 6y^{-1/3} = 3x. \tag{2.88}$$

We put

$$v = y^{-1/3},$$

then, we obtain

$$\frac{dv}{dx} = -\frac{1}{3}y^{-4/3}\frac{dy}{dx}.$$

Plugging this into (2.88), we obtain

$$-3x\frac{dv}{dx} + 6v = 3x. \tag{2.89}$$

Equation (2.90) is a first order linear equation with the depend variable v and the independent variable x, and we can rewrite it as

$$\frac{dv}{dx} - \frac{2}{x}v = -1. \tag{2.90}$$

Let

$$\mu(x) = e^{-\int \frac{2}{x}dx} = \frac{2}{x^2}.$$

Multiplying (2.90) by $\mu(x) = 2/x^2$, we obtain

$$\frac{1}{x^2}\frac{dv}{dx} - \frac{2}{x^3}v = -\frac{1}{x^2}. \tag{2.91}$$

Which gives

$$\frac{d}{dx}\left(\frac{1}{x^2}v\right) = -\frac{1}{x^2}. \tag{2.92}$$

Integrating both sides in (2.92) with respect to x, we find

$$\int \frac{d}{dx}\left(\frac{1}{x^2}v\right)dx = -\int \frac{1}{x^2}dx.$$

This yields

$$\frac{1}{x^2}v + c_1 = \frac{1}{x} + c_2.$$

Therefore, we obtain

$$v = x + cx^2, \qquad c = c_2 - c_1.$$

Since, $v = y^{-1/3}$, then we obtain

$$y(x) = \frac{1}{(x + cx^2)^3},$$

as a solution of (2.87).

2.5   Riccati Equations

Definition 2.5.1

A *Riccati equation* is a first order nonlinear differential equation that can be written in the form

$$\frac{dy}{dx} = A(x)y^2 + B(x)y + D(x), \tag{2.93}$$

where $A(x)$, $B(x)$ and $D(x)$ are continuous on an interval $I = (a,b)$.

ℹ Remark 2.5.1 We have the following particular cases:

- If $A(x) = 0$ for all x in I, then equation (2.93) reduces to a first order linear equation.
- If $D(x) = 0$ for all x in I, then equation (2.93) becomes a Bernoulli equation.

The method of solving Riccati equations If one solution $y_1(x)$ of (2.93) is known, then the general solution can be easy found by quadratures in two different ways:

1. We put the substitution

$$y(x) = y_1(x) + \frac{1}{u(x)}, \tag{2.94}$$

then, we obtain, using the chain rule

$$\frac{dy}{dx} = \frac{dy_1}{dx} - \frac{1}{u^2}\frac{du}{dx}. \tag{2.95}$$

Hence, inserting (2.95) into (2.93), we obtain

$$\frac{dy_1}{dx} - \frac{1}{u^2}\frac{du}{dx} = A(x)\left(y_1 + \frac{1}{u}\right)^2 + B(x)\left(y_1 + \frac{1}{u}\right) + D(x).$$

Since y_1 is a solution of (2.93), then the above equation leads to

$$\frac{du}{dx} + U(x)u + A(x) = 0, \tag{2.96}$$

with

$$U(x) = B(x) + 2A(x)y_1(x).$$

Equation (2.96) is a first order linear equation that can be solved using the method of integrating factor as in ▶ Sect. 2.3 to find u. Once u is found, then the general solution of (2.93) is given by (2.94).

2. We look for a general solution of the form

$$y(x) = y_1(x) + u(x). \tag{2.97}$$

Thus, the derivative of y with respect to x, leads to

$$\frac{dy}{dx} = \frac{dy_1}{dx} + \frac{du}{dx}. \tag{2.98}$$

Plugging (2.98) into (2.93) gives, by using the fact that y_1 is a solution of (2.93), the following differential equation for u:

$$\frac{du}{dx} = U(x)u + A(x)u^2, \tag{2.99}$$

which is a Bernoulli type equation that can be easily solved using the method in ▶ Sect. 2.4.2.

Example 2.17

Using the above two methods to solve the Riccati equation

$$\frac{dy}{dx} = (y - x)^2 + 1. \tag{2.100}$$

Solution

Equation (2.100) can be rewritten as

$$\frac{dy}{dx} = y^2 - 2xy + x^2 + 1, \tag{2.101}$$

which is a Riccati equation with $A(x) = 1$, $B(x) = -2x$ and $D(x) = x^2 + 1$.

It is clear that $y_1(x) = x$ is a solution of (2.101). We use the two above methods to find the general solution.

▬ **Method 1.** We look for a general solution of the form

$$y(x) = y_1(x) + \frac{1}{u(x)}$$

$$= x + \frac{1}{u(x)}$$

Thus,

$$\frac{dy}{dx} = 1 - \frac{1}{u^2}\frac{du}{dx}.$$

Plugging this into (2.101), we obtain, after a simple computation,

$$\frac{du}{dx} = -1. \tag{2.102}$$

Equation (2.102) is a first order linear equation and it is separable. It's solution is

$$u(x) = -x + c,$$

where c is a constant. Consequently, the general solution of (2.100) is

$$y(x) = x + \frac{1}{c - x}. \tag{2.103}$$

— **Method 2.** We search for a general solution of the form

$$y(x) = y_1(x) + u(x)$$
$$= x + u(x). \tag{2.104}$$

Then,

$$\frac{dy}{dx} = 1 + \frac{du}{dx}.$$

Therefore, equation (2.101) leads to

$$\frac{du}{dx} = u^2. \tag{2.105}$$

Equation (2.105) is a Bernoulli type equation and it is also separable. Using the method in ▶ Sect. 2.1, we rewrite (2.105) as

$$\frac{du}{u^2} = dx.$$

Integrating both sides in the above equation, we get

$$-\frac{1}{u} + c_1 = x + c_2.$$

Which gives

$$u(x) = \frac{1}{c - x},$$

with $c = c_1 - c_2$. Inserting this into (2.104), we deduce that the general solution of (2.100) is given by

$$y(x) = x + \frac{1}{c - x},$$

which is the same one given in (2.103).

2.5.1 Exercises

Exercise 2.13
Find the solution to the differential equation

$$x\frac{dy}{dx} = y + \sqrt{x^2 + y^2} \tag{2.106}$$

in the interval $I = \mathbb{R} - \{0\}$.

Solution
Before solving equation (2.106), we need first to find its type.
Equation (2.106) can be written as

$$\frac{dy}{dx} = \frac{y}{x} + \frac{\sqrt{x^2 + y^2}}{x} = \begin{cases} \frac{y}{x} + \sqrt{1 + (y/x)^2}, & \text{if} \quad x > 0, \\ \frac{y}{x} - \sqrt{1 + (y/x)^2}, & \text{if} \quad x < 0. \end{cases} \tag{2.107}$$

Consequently, equation (2.106) is homogeneous. Thus, we put

$$v = \frac{y}{x}.$$

Then, we find

$$\frac{dy}{dx} = v + x\frac{dv}{dx}.$$

Plugging these into (2.107), we obtain

$$v + x\frac{dv}{dx} = \begin{cases} v + \sqrt{1 + v^2}, & \text{if} \quad x > 0, \\ v - \sqrt{1 + v^2}, & \text{if} \quad x < 0. \end{cases} \qquad (2.108)$$

Which gives

$$\begin{cases} \dfrac{dv}{\sqrt{1 + v^2}} = \dfrac{dx}{x}, & \text{if} \quad x > 0, \\ \dfrac{dv}{\sqrt{1 + v^2}} = -\dfrac{dx}{x}, & \text{if} \quad x < 0. \end{cases} \qquad (2.109)$$

The equations in (2.109) are separable equations and by integrating both sides and using the fact that

$$\int \frac{dv}{\sqrt{1 + v^2}} = \sinh^{-1} v + c,$$

we obtain

$$\begin{cases} \sinh^{-1} v = \ln x + C, & \text{if} \quad x > 0, \\ \sinh^{-1} v = -\ln(-x) + C, & \text{if} \quad x < 0. \end{cases}$$

Applying the sinh function to both sides in the above formula and using the fact that $v = y/x$, we get

$$\begin{cases} y(x) = x \sinh(\ln x + C), & \text{if} \quad x > 0, \\ y(x) = x \sinh(C - \ln(-x)), & \text{if} \quad x < 0. \end{cases}$$

Consequently, the above formulas of $y(x)$ represent the explicit solutions of the differential equation (2.106).

Exercise 2.14

Find the solution to the differential equation

$$\frac{dy}{dx} + 2xy = -xy^4. \qquad (2.110)$$

Solution

Equation (2.110) is a Bernoulli type equation with $n = 4$. Thus, we put

$$v = y^{-3},$$

which gives

$$\frac{dv}{dx} = -3y^{-4}\frac{dy}{dx}. \qquad (2.111)$$

Now, we divide equation (2.110) by y^4, we obtain

$$y^{-4}\frac{dy}{dx} + 2xy^{-3} = -x. \tag{2.112}$$

Using (2.111), then (2.112) becomes

$$-\frac{1}{3}\frac{dv}{dx} + 2xv = -x. \tag{2.113}$$

Equation (2.113) is a first order linear equation for the dependent variable v. We use the method of integrating factor to find its solutions. Indeed, we first write it in the $p-q$ form as

$$\frac{dv}{dx} - 6xv = 3x, \tag{2.114}$$

with $p(x) = -6x$ and $q(x) = 3x$. We compute

$$\mu(x) = e^{\int p(x)dx} = e^{-6\int x dx}$$
$$= e^{-3x^2}.$$

Now, we multiply equation (2.114) by $\mu(x) = e^{-3x^2}$, we get

$$\frac{d}{dx}\left(e^{-3x^2}v\right) = 3xe^{-3x^2}.$$

Integrating both sides in the above equation with respect to x, we get

$$v(x) = -\frac{1}{2} + Ce^{3x^2}.$$

This implies that

$$y^{-3}(x) = -\frac{1}{2} + Ce^{3x^2},$$

is the solution to the equation (2.110).

Linear Second-Order Equations

Belkacem Said-Houari

B. Said-Houari, *Differential Equations: Methods and Applications*, Compact Textbooks in Mathematics,
DOI 10.1007/978-3-319-25735-8_3, © Springer International Publishing Switzerland 2015

We have seen in ▶ Sect. 1.1.5 that the mathematical model for the mass–spring oscillator is a linear differential equation of second order with the constant coefficients m, b and k. In this chapter we will discuss the methods of solving linear second order differential equations the form:

$$a(t)y'' + b(t)y' + c(t)y = f(t), \tag{3.1}$$

where $y = y(t)$ is the unknown (the dependent variable), $a(t), b(t)$ and $c(t)$ are continuous functions on some interval I with $a(t) \neq 0$ for t in I. The function $f(t)$ usually known as the *input* or *the forcing term* and the solution $y(t)$ is known as the *output* or *response*. As we will see later, the output $y(t)$ depends on the input $f(t)$. Hereafter, we use t as the independent variable, y the independent variable with $y' = \frac{dy}{dt}$, $y'' = \frac{d^2y}{dt^2}$ and so on. Also, without confusion, we write sometimes the functions $y, y_1, \ldots$ as $y(t), y_1(t), \ldots$

The solution of equation (3.1) depends on the solution of the equation (called *homogeneous*):

$$a(t)y'' + b(t)y' + c(t)y = 0, \tag{3.2}$$

and on the form of the term $f(t)$ on the right-hand side of (3.1). We first discuss the case where the coefficients are constants and then we investigate the case of variable coefficients.

3.1 Homogeneous Equations with Constant Coefficients

In this section, we introduce the method of solving a homogeneous second order differential equation with constant coefficients. An example of equation (3.2) is when we consider that the total external force in the mass–spring oscillator is equal to zero. That is when the mass–spring oscillator is vibrating freely. We first discuss the solution of the homogeneous equation (3.2) and after that we solve the equation (3.1).

We first give the definition of the *Wronskian* and the notion of *linearly independent* solutions.

3.1.1 Wronskian and Linear Independence

Definition 3.1.1 (Linear independence of two functions)

Let $y_1(t)$ and $y_2(t)$ be two functions defined on an interval I.
- We say that $y_1(t)$ and $y_2(t)$ are *linearly independent* on I if and only if none of them can be written as a constant multiple of the other in I. In other words, if there are two constants A and B satisfying

$$Ay_1(t) + By_2(t) = 0,$$

for all t in I, then $A = B = 0$.
- We say that $y_1(t)$ and $y_2(t)$ are *linearly dependent* if there exists a constant c such that $y_2(t) = cy_1(t)$ for all t in I.

Example 3.1
The functions $y_1(t) = e^{2t}$ and $y_2(t) = e^{3t}$ are linearly independent.

Definition 3.1.2 (Wronskian)

The *Wronskian* determinant of two functions $y_1(t)$ and $y_2(t)$ is given by

$$W[y_1, y_2](t) = \det \begin{pmatrix} y_1(t) & y_2(t) \\ y_1'(t) & y_2'(t) \end{pmatrix} = y_1(t)y_2'(t) - y_2(t)y_1'(t). \qquad (3.3)$$

Example 3.2
Find the Wronskian of the two functions $y_1(t) = \sin t$ and $y_2(t) = \cos t$.

Solution
We apply formula (3.3) to get

$$W[y_1, y_2](t) = (\sin t)(-\sin t) - (\cos t)(\cos t) = -1.$$

Lemma 3.1.1 (Linear independence and the Wronskian) If $y_1(t)$ and $y_2(t)$ are two differentiable functions on some interval I and if the Wronskian of $y_1(t)$ and $y_2(t)$ is nonzero at some point t_0 in I, then they are linearly independent.

Proof
Let t_0 in I such that $W[y_1, y_1](t_0) \neq 0$. Let A and B be two constants satisfying

$$Ay_1(t) + By_2(t) = 0 \qquad (3.4)$$

for all t in I. According to Definition 3.1.1 and in order to show that $y_1(t)$ and $y_2(t)$ are linearly independent, we need to prove that $A = B = 0$. Taking the derivative of (3.4) with respect to t, we get

$$Ay_1'(t) + By_2'(t) = 0. \qquad (3.5)$$

Collecting (3.4) and (3.5), we get the system of algebraic equations with the unknowns A and B

$$\begin{cases} Ay_1(t) + By_2(t) = 0, \\ Ay_1'(t) + By_2'(t) = 0, \end{cases}$$

which holds for all t in I. Thus, it also holds for $t = t_0$. That is

$$\begin{cases} Ay_1(t_0) + By_2(t_0) = 0, \\ Ay_1'(t_0) + By_2'(t_0) = 0. \end{cases} \tag{3.6}$$

It is clear that $A = B = 0$ is a solution to (3.6) and (from linear Algebra) since

$$W[y_1, y_2](t_0) = \det \begin{pmatrix} y_1(t_0) & y_2(t_0) \\ y_1'(t_0) & y_2'(t_0) \end{pmatrix} \neq 0,$$

then the solution $A = B = 0$ is a unique solution. Consequently $y_1(t)$ and $y_2(t)$ are linearly independent.

Example 3.3
In Example 3.2, the two functions $y_1(t) = \sin t$ and $y_2(t) = \cos t$ are linearly independent since their Wronskian

$$W[y_1, y_2](t) = -1 \neq 0.$$

Example 3.4
Show that the two functions $y_1(t) = e^{r_1 t}$ and $y_2(t) = e^{r_2 t}$ are linearly independent for any two real numbers $r_1 \neq r_2 \neq 0$.

Solution
Let us define the Wronskian of $y_1(t)$ and $y_2(t)$ as follows:

$$\begin{aligned} W[y_1, y_2](t) &= y_1(t)y_2'(t) - y_2(t)y_1'(t) \\ &= (r_2 - r_1)e^{(r_1 + r_2)t}. \end{aligned}$$

Since $r_1 \neq r_2 \neq 0$, then $W[y_1, y_2](t) \neq 0$ for all t in $\mathbb{R}$. Consequently, applying Lemma 3.1.1, we deduce that $y_1(t)$ and $y_2(t)$ are linearly independent.

Example 3.5
Let us define $y_1(t) = te^{rt}$ and $y_2(t) = e^{rt}$. Show that for any real number r, the functions $y_1(t)$ and $y_2(t)$ are linearly independent.

Solution
We compute the Wronskian of $y_1(t)$ and $y_2(t)$ as follows:

$$\begin{aligned} W[y_1, y_2](t) &= y_1(t)y_2'(t) - y_2(t)y_1'(t) \\ &= te^{rt} \cdot re^{rt} - e^{rt}(e^{rt} + rte^{rt}) \\ &= -e^{2rt} \neq 0, \end{aligned}$$

for all t. Thus, $y_1(t)$ and $y_2(t)$ are linearly independent.

Corollary 3.1.2 If $y_1(t)$ and $y_2(t)$ are linearly dependent functions, then their Wronskian must be zero at every point t.

The following theorem applies even for the second order differential equations with variable coefficients.

Theorem 3.1.3 (Abel's formula)

Let us consider the second order differential equation with variable coefficients

$$y'' + p(t)y' + q(t)y = 0, \tag{3.7}$$

where p and q are continuous on an open interval I. Let $y_1(t)$ and $y_2(t)$ be two solutions of (3.7), then, the Wronskian of $y_1(t)$ and $y_2(t)$ is:

$$W[y_1, y_2](t) = W(t) = C \exp\left\{-\int p(t)dt\right\}, \tag{3.8}$$

where C is a constant.

Proof

In order to prove the above theorem, we need just to verify that $W[y_1, y_2](t)$ satisfies the differential equation

$$\frac{dW}{dt} + p(t)W = 0. \tag{3.9}$$

Form (3.3), we have

$$\frac{dW}{dt} = y_1(t)y_2''(t) - y_2(t)y_1''(t). \tag{3.10}$$

Since $y_1(t)$ and $y_2(t)$ are solutions of (3.8), then we have

$$y_1''(t) = -p(t)y_1'(t) - q(t)y_1(t), \quad \text{and} \quad y_2''(t) = -p(t)y_2'(t) - q(t)y_2(t). \tag{3.11}$$

Inserting the equations in (3.11) into (3.10), we get

$$\frac{d}{dt}W = -p(t)y_1(t)y_2'(t) + p(t)y_2(t)y_1'(t)$$
$$= -p(t)W.$$

Thus, equation (3.9) is satisfied. This finishes the proof of Theorem 3.1.3.

3.1.2 Solutions of the Homogeneous Second Order Equation

In order to solve equation (3.2) we need to look for a *general* solution of the form

$$y(t) = c_1 y_1(t) + c_2 y_2(t), \tag{3.12}$$

where c_1 and c_2 are two positive constants and $y_1(t)$ and $y_2(t)$ are also solutions to (3.2) and are *linearly independent*. In other words, if we know two linearly independent solutions of equation (3.2). Then we will know all the solutions of the equation by taking linear combination of these two know solutions.

The form of equation (3.2) suggests that we first look to solutions of the form

$$y(t) = e^{rt},$$
(3.13)

where r is a constant. Thus, according to Example 3.4, if we found two solutions $y_1(t) = e^{r_1 t}$ and $y_2(t) = e^{r_2 t}$ with $r_1 \neq r_2 \neq 0$, then we will be able to write the general solution to the equation (3.2) as in (3.12).

Taking the first and the second derivative of (3.13) and inserting them into (3.2) (keep in mind that the coefficients are constants), we get the equation

$$e^{rt}(ar^2 + br + c) = 0,$$
(3.14)

which leads to

$$ar^2 + br + c = 0.$$
(3.15)

Therefore, $y(t) = e^{rt}$ is a solution of equation (3.2) if and only if r is a solution to the algebraic equation (3.15). Equation (3.15) is called the *characteristic* equation associated to the differential equation (3.2). We have three possibilities for the solutions of (3.15):
- The characteristic equation has tow *distinct real* roots,
- The characteristic equation has a *repeated* root,
- The characteristic equation has tow *complex conjugate* roots.

Now, we discuss the above three cases separately.

The characteristic equation has tow distinct real roots Suppose that the characteristic equation has two distinct real roots. This is the case when $\Delta = b^2 - 4ac > 0$ and the roots are

$$r_1 = \frac{-b + \sqrt{b^2 - 4ac}}{2a}, \qquad r_2 = \frac{-b - \sqrt{b^2 - 4ac}}{2a}.$$

Thus, the functions $y_1(t)$ and $y_2(t)$ are two independent solutions to the differential equation (3.2). Consequently, the general solution is given by

$$y(t) = c_1 e^{r_1 t} + c_2 e^{r_2 t},$$
(3.16)

where c_1 and c_2 are two constants.

Example 3.6
Find the solutions to the differential equation

$$y''(t) - 3y'(t) + 2y(t) = 0.$$
(3.17)

Solution

The characteristic equation associated to (3.17) is:

$$r^2 - 3r + 2 = 0.$$

This equation has tow roots: $r_1 = 1$ and $r_2 = 2$. Thus, the general solution to (3.17) is given by

$$y(t) = c_1 e^t + c_2 e^{2t}.$$

The characteristic equation has a repeated root In this case the repeated root of the characteristic equation is given by

$$r = -\frac{b}{2a}.$$

Thus, $y_1(t) = e^{rt}$ is a solution to the differential equation (3.2). Therefore, we need to find a second solution $y_2(t)$ independent to $y_1(t)$ in order to write the general solution of (3.2). There are several methods that can be used to find a second solution. We will talk about a method know as *method of reduction of order*. This method is applicable to an arbitrary second order linear homogeneous equation even with variable coefficients. Assuming that we have found one solution $y_1(t)$ to the differential equation (3.2). Since the equation is linear, then for any constant C, the function $C y_1(t)$ is also a solution. Of course we cannot take $y_2(t) = C y_1(t)$ since we are looking for two independent solutions. But if C depends on t, then the two solutions $y_1(t)$ and $y_2(t)$ can be linearly independent. This suggests that we may look for a solution of the form

$$y_2(t) = u(t) y_1(t).$$

Taking the first and the second derivatives of $y_2(t)$, we get

$$y_2'(t) = u'(t) y_1(t) + u(t) y_1'(t) \tag{3.18}$$

and

$$y_2''(t) = u''(t) y_1(t) + 2u'(t) y_1'(t) + u(t) y_1''(t). \tag{3.19}$$

For $y_2(t)$ to be a solution of (3.2), so it should satisfy the equation:

$$a y_2''(t) + b y_2'(t) + c y_2(t) = 0. \tag{3.20}$$

Inserting (3.18) and (3.19) into (3.20), we get

$$u(t)\Big(a y_1''(t) + b y_1'(t) + c y_1(t)\Big) + a u''(t) y_1(t) + u'(t)\Big(2a y_1'(t) + b y_1(t)\Big) = 0.$$

Since $y_1(t)$ is a solution, then we get

$$a u''(t) y_1(t) + u'(t)\Big(2a y_1'(t) + b y_1(t)\Big) = 0. \tag{3.21}$$

Keep in mind that $y_1(t) = e^{-\frac{b}{2a}t}$, then the second term on the left-hand side of (3.21) vanishes. Consequently, if we choose $u(t)$ such that

$$u''(t) = 0, \tag{3.22}$$

then (3.21) is satisfied. Integrating (3.22), with respect to t, we get

$$u(t) = k_1 t + k_2,$$

where k_1 and k_2 are constants. It is enough to choose just one solution. We may select $k_1 = 1$ and $k_2 = 0$, to get $u(t) = t$. Thus the second solution is

$$y_2(t) = ty_1(t) = te^{rt}.$$

According to Example 3.5, $y_1(t)$ and $y_2(t)$ are linearly independent. Consequently the general solution of the differential equation (3.2) is

$$y(t) = (c_1 + c_2 t)e^{rt}. \tag{3.23}$$

Example 3.7
Find the general solution to the differential equation:

$$y''(t) + 2y'(t) + y(t) = 0. \tag{3.24}$$

Solution
The characteristic equation associated to (3.24) is:

$$r^2 + 2r + 1 = 0.$$

This equation has $r = -1$ as a repeated root. Thus, the general solution of (3.24) is given by

$$y(t) = (c_1 + c_2 t)e^{-t}.$$

The characteristic equation has tow complex conjugate roots Here we assume that the characteristic equation (3.15) has two complex conjugate roots:

$$r_1 = \alpha + i\beta, \qquad r_2 = \alpha - i\beta,$$

where

$$\alpha = -\frac{b}{2a}, \qquad \beta = \frac{\sqrt{4ac - b^2}}{2a}.$$

In this case, using the Euler formula

$$e^{i\theta} = \cos\theta + i\sin\theta, \tag{3.25}$$

we get

$$e^{r_1 t} = e^{\alpha t}(\cos\beta t + i\sin\beta t), \qquad e^{r_2 t} = e^{\alpha t}(\cos\beta t - i\sin\beta t).$$

Now, we have the following lemma:

Lemma 3.1.4 Let $z(t) = u(t) + iv(t)$ be a solution of the differential equation (3.2), where the coefficients a, b and c are real numbers. Then the real part $u(t)$ and the imaginary part $v(t)$ are real-valued solutions of (3.2).

Proof

Since $z(t)$ is a solution to (3.2), then we have

$$az''(t) + bz'(t) + cz(t) = 0.$$

This implies,

$$au''(t) + bu'(t) + cu(t) + i(av''(t) + bv'(t) + cv(t)) = 0,$$

This leads to

$$au''(t) + bu'(t) + cu(t) = 0 \quad \text{and} \quad av''(t) + bv(t)' + cv(t) = 0.$$

This means that $u(t)$ and $v(t)$ are both solutions to (3.2). This completes the proof of Lemma 3.1.4.

Now, using the above lemma, we deduce that $e^{\alpha t} \cos \beta t$ and $e^{\alpha t} \sin \beta t$ are two linearly independent solutions to (3.2). Consequently, the general solution in this case is given by

$$y(t) = c_1 e^{\alpha t} \cos \beta t + c_2 e^{\alpha t} \sin \beta t. \tag{3.26}$$

Remark 3.1.5 In the case of a complex roots, the solution may also be written as

$$y(t) = C_1 e^{(\alpha+i\beta)t} + C_2 e^{(\alpha-i\beta)t}, \tag{3.27}$$

where C_1 and C_2 are complex numbers. Since $y(t)$ is a real solution, then $\bar{y}(t) = y(t)$, which gives $C_2 = \bar{C}_1$. Consequently, formula (3.27) can be rewritten as

$$y(t) = (c + id)e^{(\alpha+i\beta)t} + (c - icd)e^{(\alpha-i\beta)t}, \tag{3.28}$$

where c and d are real numbers.

We can simply recover the formula (3.26) from (3.28) by using the fact that

$$\cos \beta t = \frac{e^{i\beta t} + e^{-i\beta t}}{2}, \qquad \sin \beta t = \frac{e^{i\beta t} - e^{-i\beta t}}{2i}.$$

Example 3.8

Find the solution to the differential equation

$$y''(t) - 2y'(t) + 2y(t) = 0. \tag{3.29}$$

Solution

The characteristic equation associated to (3.29) is:

$$r^2 - 2r + 2 = 0.$$

The roots of the above equation are:

$$r_1 = 1 + i, \qquad r_2 = 1 - i.$$

Consequently, the general solution of (3.29) is given by

$$y(t) = e^t(c_1 \cos t + c_2 \sin t).$$

Example 3.9

Find the solution of second order equation

$$y''(t) - 2\alpha y'(t) + y(t) = 0, \tag{3.30}$$

where α is a real number.

Solution

The characteristic equation associated to (3.30) is

$$r^2 - 2\alpha r + 1 = 0. \tag{3.31}$$

Its discriminant is $\Delta = \alpha^2 - 1$. So, we have the following three cases:

1. If $\alpha^2 - 1 > 0$, which means that α in $(-\infty, -1) \cup (1, +\infty)$, then (3.31) has two real distinct roots

$$r_1 = \alpha + \sqrt{\alpha^2 - 1}, \qquad r_2 = \alpha - \sqrt{\alpha^2 - 1}.$$

Consequently, the solution of (3.30) is

$$y(t) = C_1 e^{(\alpha+\sqrt{\alpha^2-1})t} + C_2 e^{(\alpha-\sqrt{\alpha^2-1})t},$$

where C_1 and C_2 are two constants.

2. If $\alpha^2 = 1$, which means if $\alpha = \pm 1$, then equation (3.31) has one repeated root

$$r_0 = \alpha.$$

Therefore, the solution of (3.30) is

$$y(t) = (C_1 + C_2 t)e^{\alpha t}.$$

3. If $\alpha^2 - 1 < 0$, that is if α in $(-1, 1)$, then equation (3.31) has two complex conjugate solutions

$$r_1 = \alpha + i\sqrt{1 - \alpha^2}, \qquad r_2 = \alpha - i\sqrt{1 - \alpha^2}.$$

In this case the solution of (3.30) is

$$y(t) = e^{\alpha t}\left(C_1 \cos(\sqrt{1-\alpha^2}\ t) + C_2 \sin(\sqrt{1-\alpha^2}\ t)\right).$$

3.1.3 Harmonic Oscillators

In this subsection, we provide more discussion of the case of complex roots for the characteristic equation associated to the homogeneous second order equation with constant coefficients.

Definition 3.1.3

A harmonic oscillator is a system that, when displaced from its equilibrium position experiences a restoring force F proportional to the displacement y: $F = -ky$, where k is a positive constant.

Examples includes pendulum, masses connected to springs (as in ▶ Sect. 1.1.5), acoustic systems and electrical harmonic oscillators such as RLC circuit (an electrical circuit consisting of a resistor, an inductor, and a capacitor, connected in series or in parallel).

In ▶ Sect. 1.1.5, if we assume that the external forces are identically zero, then we get the following linear second order equation with constant coefficients:

$$my'' + by' + ky = 0. \tag{3.32}$$

Equation (3.32) can be written in the form

$$y'' + 2py' + \omega_0^2 y = 0, \tag{3.33}$$

with $p = \frac{b}{2m}$ and $\omega_0 = \sqrt{k/m}$ is the *circular frequency*. The characteristic equation associated to (3.33) is:

$$r^2 + 2pr + \omega_0^2 = 0. \tag{3.34}$$

3.1.4 Undamped Oscillators

The case $p = 0$ corresponds to the undamped case and then the solution of (3.33) is given by

$$y(t) = c \cos \omega_0 t + d \sin \omega_0 t, \tag{3.35}$$

where c and d are two positive constants.

Suppose that we choose the initial time $t_0 = 0$ and we give the initial position $y(0) = y_0$ and the initial velocity $y'(0) = v_0$. Then, by replacing t with 0 in formula (3.35) and its derivative with respect to t, we easily obtain that $c = y_0$ and $d = v_0/\omega$. In this case the solution (3.35) can be rewritten as

$$y(t) = y_0 \cos \omega_0 t + \frac{v_0}{\omega} \sin \omega_0 t. \tag{3.36}$$

The solution (3.35) can be also written as

$$y(t) = A \cos(\omega_0 t - \phi), \tag{3.37}$$

with

$$A = \sqrt{c^2 + d^2}, \quad c = A \cos \phi, \quad d = A \sin \phi, \quad \tan \phi = \frac{d}{c}.$$

The graph of the solution (3.37) is a displaced cosine wave that describes a periodic motion of the mass m. The positive constant A represents the *amplitude* of the motion and ϕ is the *phase lag* and measures the displacement of the wave from its normal position corresponding to $\phi = 0$. The *period* of the motion is $T = \frac{2\pi}{\omega_0} = 2\pi \sqrt{m/k}$. See ◼ Fig. 3.1.

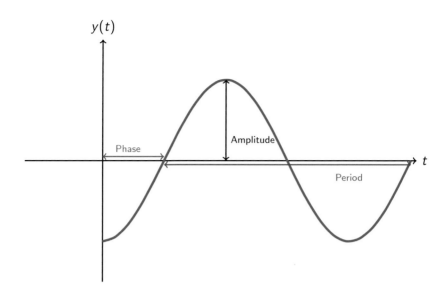

Fig. 3.1 Undamped harmonic oscillator

The total energy of the system The total energy $E(t)$ of the harmonic oscillator is the sum of the potential energy

$$V(t) = \frac{1}{2}ky^2(t)$$

and the kinetic energy

$$U(t) = \frac{m}{2}(y'(t))^2.$$

That is

$$
\begin{aligned}
E(t) &= \frac{1}{2}ky^2(t) + \frac{m}{2}(y'(t))^2 \\
&= \frac{1}{2}k\left(y_0 \cos \omega_0 t + \frac{v_0}{\omega} \sin \omega_0 t\right)^2 \\
&\quad + \frac{m}{2}\left(-\omega_0 y_0 \sin \omega_0 t + v_0 \cos \omega_0 t\right)^2
\end{aligned}
\tag{3.38}
$$

By keeping in mind that $\omega_0 = \sqrt{k/m}$, we get from above that

$$E(t) = \frac{m}{2}v_0^2 + \frac{k}{2}y_0^2 = E(0). \tag{3.39}$$

Formula (3.39) shows that the total energy in the undamped oscillator is conserved and the initial data y_0 and v_0 determine directly the amount of the total energy.

■ **Fig. 3.2** Energy diagram

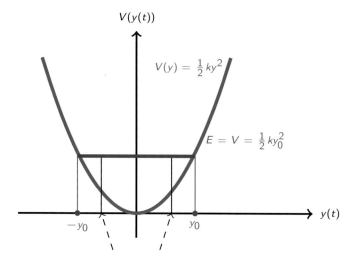

at any time $t : E(t) = V(t) + U(t)$

Energy diagram The mass starts at $y(0) = y_0$ with the potential energy only ($E(0) = V(0)$, $U(0) = 0$). The force of the spring accelerates it towards the equilibrium $y = 0$ by converting the potential energy into a kinetic energy. At the equilibrium all the energy is in the form of kinetic energy ($E(t) = U(t)$, $V(t) = 0$). Even thought the mass is force-free, at this point, it moves through to load the spring converting the kinetic energy back into potential energy. As in ■ Fig. 3.2.

3.1.5 The Damped Oscillators

For $p \neq 0$, the solutions of (3.33) are given by

$$r_1 = -p + \sqrt{p^2 - \omega_0^2}, \qquad r_2 = -p - \sqrt{p^2 - \omega_0^2}. \tag{3.40}$$

▬ For $\omega_0 > p_0$, this corresponds to the *weakly damped* oscillator. The roots are complex and the solution is given by

$$y(t) = e^{-pt} \left(c \cos \sqrt{(\omega_0^2 - p^2)}\, t + d \sin \sqrt{(\omega_0^2 - p^2)}\, t \right).$$

In this case the frequency ω_1 is smaller than the frequency ω_0 related to the undamped case. See ■ Fig. 3.3.

▬ The case $\omega_0 < p$ know as the the *strongly damped* oscillator, in this case the solution of (3.34) are real and the solution of the differential equation (3.33) is given by

$$y(t) = c \exp\left[-p + \sqrt{p^2 - \omega_0^2} \right] t + d \exp\left[-p - \sqrt{p^2 - \omega_0^2} \right] t.$$

This represents a non oscillating motion and it is a aperiodic motion. See ■ Fig. 3.4.

Fig. 3.3 Weakly damped oscillators

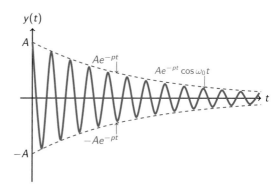

Fig. 3.4 Strongly damped oscillators

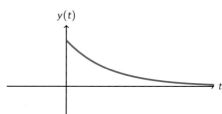

Fig. 3.5 Critically damped oscillators

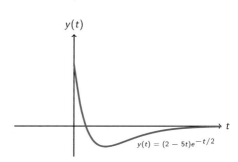

- For $p = \omega_0$, we are in a situation of *critical damped* oscillator, and the solution is given by

$$y(t) = (c + dt)e^{-pt},$$

which is illustrated in ▪ Fig. 3.5.

The energy for $p \neq 0$ In this case we multiply equation (3.32) by $y'(t)$, then, we deduce that the energy $E(t)$ defined in (3.38) satisfies

$$\frac{d}{dt}E(t) = -b(y'(t))^2, \tag{3.41}$$

since $b > 0$, then the total energy of the system is decreasing with respect to time. This means that the oscillator is loosing the energy due to the damping. In addition, the identity (3.41) shows that the loss in the energy is higher when the mass moves faster.

3.1.6 The Euler Equation

In this subsection, we discuss the solutions of a second order linear differential equation with variable coefficients known as Euler's equation. This equation has the following form

$$t^2 y''(t) + \alpha t y'(t) + \beta y(t) = 0, \qquad t > 0, \tag{3.42}$$

where α and β are two constants.

In order to solve equation (3.42), we want to turn it into a second order differential equation with constant coefficients. We introduce the change of variables

$$t = e^\tau, \qquad \text{and} \qquad Y(\tau) = y(t). \tag{3.43}$$

Thus, we have by using the chain rule and the fact that $\tau = \ln t$,

$$\frac{dy}{dt} = \frac{dY}{d\tau}\frac{d\tau}{dt} = \frac{1}{t}\frac{dY}{d\tau} \tag{3.44}$$

and

$$\frac{d^2 y}{dt^2} = \frac{1}{t^2}\frac{d^2 Y}{d\tau^2} - \frac{1}{t^2}\frac{dY}{d\tau}. \tag{3.45}$$

Plugging (3.44) and (3.45) into (3.42), we get

$$\frac{d^2 Y(\tau)}{d\tau^2} + (\alpha - 1)\frac{dY(\tau)}{d\tau} + \beta Y(\tau) = 0. \tag{3.46}$$

Equation (3.46) is a second order linear differential equation with constant coefficients where the independent variable is τ and the dependent variable is Y. As we have seen in ▶ Sect. 3.1.2, its solution depends on the roots of the characteristic equation

$$r^2 + (\alpha - 1)r + \beta = 0. \tag{3.47}$$

Thus, we have the following three possibilities:
- If the roots r_1 and r_2 of (3.47) are real distinct, then the general solution of (3.46) is given by

$$Y(\tau) = C_1 e^{r_1 \tau} + C_2 e^{r_2 \tau}.$$

This, implies, by putting $\tau = \ln t$, that the solution of (3.42) is given by

$$y(t) = C_1 t^{r_1} + C_2 t^{r_2}. \tag{3.48}$$

- If equation (3.47) has a multiple root r, then the general solution of (3.46) has the form

$$Y(\tau) = (C_1 + C_2 \tau)e^{r\tau}.$$

Thus, the solution of (3.42) is

$$y(t) = (C_1 + C_2 \ln t)t^r. \tag{3.49}$$

- If equation (3.47) has tow complex roots $r_1 = \lambda + i\mu$ and $r_2 = \lambda - i\mu$, then the general solution of (3.46) has the form

$$Y(\tau) = e^{\lambda\tau}(C_1 \cos \mu\tau + C_2 \sin \mu\tau).$$

Consequently, in this case the solution of (3.42) takes the form

$$y(t) = t^{\lambda}\Big(C_1 \cos(\mu \ln t) + C_2 \sin(\mu \ln t)\Big). \tag{3.50}$$

Example 3.10

Find the solutions to the differential equation

$$t^2 y''(t) + 3t y'(t) + 4y(t) = 0, \tag{3.51}$$

in the interval $I = (0, \infty)$.

Solution

Equation (3.51) is an Euler type equation with $\alpha = 3$ and $\beta = 4$. Thus the characteristic equation (3.47) becomes

$$r^2 + 2r + 4 = 0$$

and has the following complex roots:

$$r_1 = -1 + \sqrt{3}i, \qquad r_2 = -1 - \sqrt{3}i.$$

Consequently, the general solution of (3.51) is given by

$$y(t) = t^{-1}\Big(C_1 \cos(\sqrt{3} \ln t) + C_2 \sin(\sqrt{3}t)\Big),$$

where C_1 and C_2 are positive constants.

3.2 Homogeneous Equations with Variable Coefficients

Our aim in this section is to look for a general solution of equation (3.2) where $a(t), b(t)$ and $c(t)$ are not necessarily constants.

3.2.1 Method of Reduction of Order

In order to introduce this method, let us first write equation (3.2) in the standard form

$$y'' + p(t)y' + q(t)y = 0, \tag{3.52}$$

where $p(t) = b(t)/a(t)$ and $q(t) = c(t)/a(t)$. As, we have said before, if we can find two independent solutions $y_1(t)$ and $y_2(t)$ of (3.52), then the general solution $y(t)$ is a linear combination of these two solutions. That is:

$$y(t) = C_1 y_1(t) + C_2 y_2(t). \tag{3.53}$$

The method of *reduction of order* allows us to find a second solution to the homogeneous equation (3.52) if we already know one solution. In general this method can be applied to any linear differential equation of nth order.

The basic idea of the method is the following: Assume that we know one solution $y_1(t)$ to the differential equation (3.52) and we want to find a second solution $y_2(t)$ linearly indepen-dent to $y_1(t)$, then inspired by the method used in ▶ Sect. 3.1.2 for the repeated root case, we may look for a second solution in the form

$$y_2(t) = u(t)y_1(t). \tag{3.54}$$

Taking the first and the second derivatives of (3.54) with respect to t, we get

$$y_2'(t) = u'(t)y_1(t) + u(t)y_1'(t) \tag{3.55}$$

and

$$y_2''(t) = u''(t)y_1(t) + 2u'(t)y_1'(t) + u(t)y_1''(t). \tag{3.56}$$

Since we claim that $y_2(t)$ is a solution to (3.52), then we have

$$y_2'' + p(t)y_2' + q(t)y_2 = 0. \tag{3.57}$$

Now, inserting (3.55) and (3.56) into (3.57), we get

$$u''(t)y_1(t) + 2u'(t)y_1'(t) + p(t)u'(t)y_1(t) + u(t)\left[y_1''(t) + p(t)y_1'(t) + q(t)y_1(t)\right] = 0.$$

Since $y_1(t)$ is a solution, then the above equation leads to

$$y_1(t)u''(t) + \left(2y_1'(t) + p(t)y_1(t)\right)u'(t) = 0. \tag{3.58}$$

It is obvious that equation (3.58) is a linear equation for the independent variable $u'(t)$. We may easily solve it by the substitution $z = u'(t)$. Then we get

$$y_1(t)z'(t) + \left(2y_1'(t) + p(t)y_1(t)\right)z(t) = 0. \tag{3.59}$$

Consequently, equation (3.59) can be also seen as first order equation of separable variables. Using the method in ▶ Sect. 2.1, we find that

$$\frac{1}{z(t)}\frac{dz(t)}{dt} = -\frac{2y_1'(t) + p(t)y_1(t)}{y_1(t)} = -2\frac{y_1'(t)}{y_1(t)} - p(t).$$

This gives

$$\frac{dz(t)}{z(t)} = \left[-2\frac{y_1'(t)}{y_1(t)} - p(t)\right]dt. \tag{3.60}$$

Integrating both sides in (3.60) and applying the exponential to the result, we get that

$$z(t) = \frac{1}{y_1^2(t)}\exp\left\{\int -p(t)dt\right\} \tag{3.61}$$

is one of the solutions. Keeping in mind that $u'(t) = z(t)$, we obtain

$$u(t) = \int \left(\frac{1}{y_1^2(t)} \exp \left\{ \int -p(t)dt \right\} \right) dt. \tag{3.62}$$

Consequently, the second solution $y_2(t)$ is

$$y_2(t) = y_1(t) \int \left(\frac{1}{y_1^2(t)} \exp \left\{ \int -p(t)dt \right\} \right) dt. \tag{3.63}$$

Next, we need to show that $y_1(t)$ and $y_2(t)$ are linearly independent.

ⓘ Lemma 3.2.1 The solution

$$y_2(t) = y_1(t) \int \left(\frac{1}{y_1^2(t)} \exp \left\{ \int -p(t)dt \right\} \right) dt, \tag{3.64}$$

of equation (3.52) is linearly independent with $y_1(t)$.

Proof
Assume that $y_1(t)$ and $y_2(t)$ are linearly dependent. Then, there exists a constant C such that $y_2(t) = Cy_1(t)$. On the other hand, from (3.64), we have

$$\frac{d}{dt} \left(\frac{y_2(t)}{y_1(t)} \right) = \frac{1}{y_1^2(t)} \exp \left\{ \int -p(t)dt \right\} = 0,$$

since $y_2(t)/y_1(t) = C$. This is a contradiction since $\exp \left\{ \int -p(t)dt \right\} \neq 0$. Consequently, $y_1(t)$ and $y_2(t)$ are linearly independent.

Example 3.11
Find the general solution to the differential equation

$$y''(t) \tan t + y'(t)(\tan^2 t - 2) + y(t) \frac{2}{\tan t} = 0, \tag{3.65}$$

for t in $(0, \pi/2)$, provided that $y_1(t) = \sin t$ is a solution.

Solution
Since $y_1(t) = \sin t$ is a solution to (3.65), then we need to search for a second solution $y_2(t)$ independent to $y_1(t)$ of the form

$$y_2(t) = u(t)y_1(t) = u(t) \sin t. \tag{3.66}$$

Consequently,

$$y_2'(t) = u'(t) \sin t + u(t) \cos t, \quad y_2'' = u''(t) \sin t + 2u'(t) \cos t - u(t) \sin t. \tag{3.67}$$

Plugging (3.66) and (3.67) into (3.65), we find

$$\left(u''(t) \sin t + 2u'(t) \cos t - u(t) \sin t \right) \tan t$$
$$+ \left(u'(t) \sin t + u(t) \cos t \right)(\tan^2 t - 2) + u(t) \sin t \frac{2}{\tan t} = 0. \tag{3.68}$$

3

Since $y_1(t) = \sin t$ is a solution to (3.65), then we have

$$u(t)\left(-\sin t \cdot \tan t + \cos t \left((\tan^2 t - 2)\right) + \sin t \frac{2}{\tan t}\right) = 0.$$

Consequently, we get for all $t \in (0, \pi/2)$,

$$u'' + u' \tan t = 0. \tag{3.69}$$

From equation (3.69), we get

$$\frac{u''(t)}{u'(t)} = -\tan t.$$

Integrating the above equation with respect to t, we obtain

$$\ln |u'(t)| = \ln |\cos t| + c_1.$$

Taking the exponential in both sides in the above equation, we obtain

$$u'(t) = c_2 \cos t, \tag{3.70}$$

where $c_2 = \pm e^{c_1}$. Integrating (3.70) with respect to t, we find

$$u(t) = c_2 \sin t + c_3.$$

Keeping in mind (3.66), we finally get

$$y_2(t) = c_2 \sin^2 t + c_3 \sin t, \tag{3.71}$$

which is the second solution of (3.65).

The method of reduction of order by using Abel's formula Suppose that we know one solution $y_1(t)$ of the equation

$$y'' + p(t)y' + q(t)y = 0. \tag{3.72}$$

To find the second solution $y_2(t)$, we do the following:
- Compute the Wronskian using Abel's formula:

$$W[y_1, y_2](t) = C \exp\left\{-\int p(t)dt\right\}. \tag{3.73}$$

- Calculate the Wronskian directly using (3.3), that is

$$W[y_1, y_2](t) = y_1(t)y_2'(t) - y_2(t)y_1'(t). \tag{3.74}$$

- Set the two expression (3.73) and (3.74) equal. Then, the result will be a first order differential equation for $y_2(t)$.
- Solve the differential equation for $y_2(t)$.

Example 3.12
Find the second solution in Example 3.11 using Abel's formula.

Solution
Let us apply the above method to equation (3.65). Indeed, we first write equation (3.65) as

$$y''(t) + y'(t)\frac{\tan^2 t - 2}{\tan t} + y(t)\frac{2}{\tan^2 t} = 0. \tag{3.75}$$

We compute

$$W[y_1, y_2](t) = C \exp\left\{-\int\left(\tan t - \frac{2}{\tan t}\right)dt\right\}$$
$$= C \exp\{\ln |\cos t \sin^2 t|\}$$
$$= C \cos t \sin^2 t, \tag{3.76}$$

since $t \in (0, \pi/2)$.

On the other hand from (3.74), we have

$$y_1(t)y_2'(t) - y_2(t)y_1'(t) = y_2' \sin t - y_2(t)\cos t. \tag{3.77}$$

Consequently, we have from (3.76) and (3.77)

$$y_2' \sin t - y_2(t)\cos t = C \cos t \sin^2 t,$$

which is equivalent to

$$y_2' - y_2(t)\frac{1}{\tan t} = C \sin t \cos t. \tag{3.78}$$

Equation (3.78) is a linear first order equation, then, we may solve it using the method of integrating factor (see ▶ Sect. 2.3). We compute

$$\mu(t) = \exp\left\{-\int \frac{dt}{\tan t}\right\} = \frac{1}{\sin t}.$$

Multiplying (3.78) by $\mu(t) = 1/\sin t$, we find

$$\frac{d}{dt}\left\{\frac{y_2(t)}{\sin t}\right\} = C \cos t.$$

Integrating both sides in the above equation, we get

$$y_2(t) = C \sin^2 t + K \sin t.$$

where C and K are constants. Thus, we obtained the same formula (3.71) for $y_2(t)$.

3.3   Non-Homogeneous Linear Second Order Equations

Our goal in this section is to solve the non-homogeneous second order differential equation

$$a(t)y'' + b(t)y' + c(t)y = f(t), \tag{3.79}$$

where $f \neq 0$. This is the case for instance where the sum of the external forces in the spring-mass model $F_e = f(t)$ (▶ Sect. 1.1.5) is not identically zero. For example an external force can be a magnetic field (acting on the steel mass) or vibration of the wall. See also the phenomenon of resonance in ▶ Sect. 3.3.2.

We assume that the general solution of the corresponding homogeneous equation

$$a(t)y'' + b(t)y' + c(t)y = 0, \tag{3.80}$$

is already known. We begin by proving the following theorem.

Theorem 3.3.1

We consider the second order linear differential equation

$$a(t)y'' + b(t)y' + c(t)y = f(t), \tag{3.81}$$

where $a(t)$, $b(t)$ and $c(t)$ are continuous functions on some interval I with $a(t) \neq 0$ for t in I. Then, the solution of (3.81) is given by

$$y(t) = c_1 y_1(t) + c y_2(t) + y_p(t), \tag{3.82}$$

where $y_1(t)$ and $y_2(t)$ are two linearly independent solutions of the homogeneous equation

$$a(t)y'' + b(t)y' + c(t)y = 0, \tag{3.83}$$

and $y_p(t)$ is a particular solution of (3.81) and c_1 and c_2 are arbitrary constants.

Proof

We can write equation (3.81) as

$$L[y(t)] = f(t),$$

where L is the differential operator

$$L = a(t)D^2 + b(t)D + c(t)$$

and $D^2 = \frac{d^2}{dt^2}$, $D = \frac{d}{dt}$. Since L is a linear operator, then we have

$$L[c_1 y_1(t) + c_2 y_2(t) + y_p(t)] = c_1 L[y_1(t)] + c_2 L[y_2(t)] + L[y_p(t)].$$

Since $y_1(t)$ and $y_2(t)$ are solutions to the homogeneous equation (3.83), then $L[y_1(t)] = L[y_2(t)] = 0$. On the other hand, since $y_p(t)$ is a particular solution to (3.81), then we have $L[y_p(t)] = f(t)$. Consequently, we get $L[y(t)] = f(t)$. Thus, $y(t)$ is a solution to the differential equation (3.81). Thus, we proved that any function $y(t)$ written in the form (3.82) is a solution to the differential equation (3.81).

Next, we need to show that any other solution to (3.81) is in the form (3.82). Indeed, let $u(t)$ be another solution to (3.81), thus, we have

$$L[u(t)] = f(t).$$

On the other hand, since $y_p(t)$ is a solution to the equation (3.81), we have

$$L[y_p(t)] = f(t).$$

Consequently, we have from the above two equations and since L is a linear operator,

$$L[u(t) - y_p(t)] = 0.$$

Therefore, $u(t) - y_p(t)$ is a solution to the homogeneous equation (3.83). Thus, there exist two constants $\tilde{c}_1$ and $\tilde{c}_2$ such that

$$u(t) - y_p(t) = \tilde{c}_1 y_1(t) + \tilde{c}_2 y_2(t),$$

since any solution to (3.83) is a linear combination of $y_1(t)$ and $y_2(t)$. Consequently, $u(t)$ is written in the form (3.82).

Remark 3.3.2 Theorem 3.3.1 has its root in the first order equation studied in ▶ Sect. 2.3. We have seen that the solution of the linear first order differential equation

$$\frac{dy}{dt} + p(t)y = q(t), \tag{3.84}$$

is written in the form

$$y(t) = \frac{1}{\mu(t)} \left(\int \mu(t)q(t)dt + C \right)$$
$$= \underbrace{e^{-\int p(t)dt} \left[\int \left(e^{\int p(t)dt} \right) q(t)dt \right]}_{:= y_p(t)} + Ce^{-\int p(t)dt} \tag{3.85}$$

So, it can be easily seen that $y(t) = e^{-\int p(t)dt}$ is the solution of the equation

$$\frac{dy}{dt} + p(t)y = 0, \tag{3.86}$$

and $y_p(t)$ is a particular solution of (3.84).

3.3.1 Non-Homogeneous Equations with Constant Coefficients

In this subsection, we are going to show some methods of finding solutions to the differential equation (3.1) where $a(t)$, $b(t)$ and $c(t)$ are constants.

Methods of finding particular solutions Now, we consider the second order linear differential equation with constant coefficients:

$$ay'' + by' + cy = f(t).$$ (3.87)

Our goal is to find a particular solution $y_p(t)$ of (3.87) in the case where the input function $f(t)$ has an exponential form. This includes the following examples: $f(t) = e^{\alpha t}$, $f(t) = \sin \beta t$, $f(t) = \cos \beta t$, $f(t) = e^{\alpha t} \cos \beta t$, $f(t) = e^{\alpha t} \sin \beta t$. All these are particular case of the function $f(t) = e^{\theta t}$, where θ is a complex number.

We define the differential operator

$$P(D) = aD^2 + bD + c.$$

Then since P is linear, then, it is easy to show that

$$P(D)[e^{\theta t}] = P(\theta)e^{\theta t}.$$ (3.88)

Now, we have the following theorem.

Theorem 3.3.3
Assume that

$$P(\theta) \neq 0.$$ (3.89)

Then, the particular solution to the differential equation

$$ay''(t) + by'(t) + cy(t) = e^{\theta t},$$ (3.90)

where a, b, c are constants and θ is a complex number, is given by

$$y_p(t) = \frac{e^{\theta t}}{P(\theta)}.$$ (3.91)

Proof
To show that $y_p(t)$ defined by (3.91) is a particular solution to (3.90), then it is enough to prove that

$$P(D)[y_p(t)] = e^{\theta t}.$$

Indeed, we have

$$P(D)[y_p(t)] = P(D)\left[\frac{e^{\theta t}}{P(\theta)}\right] = \frac{1}{P(\theta)} P(D)[e^{\theta t}]$$
$$= e^{\theta t},$$

where we have used (3.88).

Example 3.13

Find a particular solution to the linear second order differential equation

$$y''(t) - y'(t) + 2y(t) = 10e^{-t} \sin t. \tag{3.92}$$

Solution

The particular solution $y_p(t)$ of (3.92) is the imaginary part of the particular solution $\tilde{y}_p(t)$ of the equation

$$(D^2 - D + 2)\tilde{y}_p(t) = 10e^{(-1+i)t}. \tag{3.93}$$

Applying formula (3.91), we obtain

$$\tilde{y}_p(t) = \frac{10e^{(-1+i)t}}{(-1+i)^2 - (-1+i) + 2}$$
$$= \frac{5}{3}(1+i)e^{-t}(\cos t + i \sin t),$$

where we have used the Euler formula (3.25). Consequently,

$$y_p(t) = \Im(\tilde{y}_p(t)) = \frac{5}{3}e^{-t}(\cos t + \sin t).$$

Example 3.14

Find a particular solution to the differential equation

$$y''(t) + 3y'(t) + 2y(t) = \sin t. \tag{3.94}$$

Solution

We first write equation (3.14) in the form

$$(D^2 + 3D + 2)y(t) = \sin t. \tag{3.95}$$

Thus, the particular solution $y_p(t)$ of (3.94) is the imaginary part of the solution $\tilde{y}_p(t)$ of the equation

$$(D^2 + 3D + 2)\tilde{y}_p(t) = e^{it}. \tag{3.96}$$

Using formula (3.91), we have

$$\tilde{y}_p(t) = \frac{e^{it}}{i^2 + 3i + 2}$$
$$= \frac{(1 - 3i)e^{it}}{10}$$
$$= \frac{1}{10}(1 - 3i)(\cos t + i \sin t).$$

Since $y_p(t) = \Im(\tilde{y}_p(t))$, then we get

$$y_p(t) = \frac{1}{10} \sin t - \frac{3}{10} \cos t.$$

As, we have seen above, Theorem 3.3.3 can be used only in the case where $P(\theta) \neq 0$. So, the natural question is: *how can we find a particular solution to* (3.90) *if* $P(\theta) = 0$? To handle this case, we start with the following lemma know as the *exponential shift rule*.

ⓘ Lemma 3.3.4 (Exponential shift) Let $P(D)$ be a polynomial operator of D. Then, for any sufficiently differentiable function $u(t)$ and for any complex number θ we have

$$P(D)[e^{\theta t} u(t)] = e^{\theta t} P(D + \theta)[u(t)]. \tag{3.97}$$

Proof
We prove (3.97) for $P(D) = D$ and $P(D) = D^2$ and then it can be easily generalized to any polynomial operator of D by induction.
We have for $P(D) = D$:

$$D[e^{\theta t} u(t)] = D[u(t)]e^{\theta t} + \theta e^{\theta t} u(t)$$
$$= e^{\theta t}(D + \theta)[u(t)].$$

Now, for $P(D) = D^2$, we have

$$D^2[e^{\theta t} u(t)] = D[e^{\theta t}(D + \theta)[u(t)]]$$
$$= e^{\theta t}(D + \theta)[(D + \theta)u(t)]$$
$$= e^{\theta t}(D + \theta)^2[u(t)]$$
$$= e^{\theta t} P(D + \theta)[u(t)],$$

which is (3.97).

Now, we state and prove the following theorem of finding a particular solution to the differential equation (3.90), when $P(\theta) = 0$.

Theorem 3.3.5
Let $P(D) = aD^2 + bD + c$. We consider the following differential equation

$$P(D)[y(t)] = e^{\theta t}. \tag{3.98}$$

Assume that $P(\theta) = 0$, then we have:
(i) if $P'(\theta) \neq 0$, meaning that if θ is a simple root of $P(D)$, then the particular solution of (3.98) is given by

$$y_p(t) = \frac{t e^{\theta t}}{P'(\theta)}. \tag{3.99}$$

(ii) If $P'(\theta) = 0$, meaning that if θ is a repeated root of $P(D)$, then the particular solution of (3.98) is given by

$$y_p(t) = \frac{t^2 e^{\theta t}}{P''(\theta)}. \tag{3.100}$$

Proof

Let us first prove (3.99). Since θ is a simple root of $P(D)$, then we may write $P(\theta)$ in the form

$$P(D) = a(D - \theta)(D - \eta),\tag{3.101}$$

where η is a complex number such that $\eta \neq \theta$. It is clear from the above formula that

$$P'(D) = a(D - \theta) + a(D - \eta).$$

Thus,

$$P'(\theta) = a(\theta - \eta).\tag{3.102}$$

Now, using formula (3.97), we obtain

$$\begin{aligned}
P(D)\left[\frac{e^{\theta t} \cdot t}{P'(\theta)}\right] &= \frac{1}{P'(\theta)}e^{\theta t} P(D + \theta)[t] \\
&= \frac{1}{P'(\theta)}e^{\theta t} a(D + \theta - \eta)D[t] \\
&= \frac{1}{a(\theta - \eta)}e^{\theta t} a(\theta - \eta) \\
&= e^{\theta t},
\end{aligned}$$

where we have used the formulas (3.101) and (3.102).

By the same method, we may prove (3.100). Indeed, if θ is a repeated root of $P(D)$, then we have

$$P(D) = a(D - \theta)^2.$$

This gives

$$P'(D) = 2a(D - \theta), \quad \text{and} \quad P''(D) = 2a.$$

Now, we have

$$\begin{aligned}
P(D)\left[\frac{t^2 e^{\theta t}}{P''(\theta)}\right] &= \frac{1}{P''(\theta)}e^{\theta t} P(D + \theta)[t^2] \\
&= \frac{1}{2a}e^{\theta t} a(D)[t^2] \\
&= e^{\theta t}.
\end{aligned}$$

This completes the proof of Theorem 3.3.5.

Example 3.15

Find a particular solution to the differential equation

$$y''(t) - 3y'(t) + 2y(t) = e^t.\tag{3.103}$$

Solution

It is clear that $\theta = 1$ is a simple root to $P(D) = D^2 - 3D + 2$. Thus, applying formula (3.100), we get

$$\begin{aligned}
y_p(t) &= \frac{te^{\theta t}}{P'(\theta)} \\
&= \frac{te^{\theta t}}{P'(1)} \\
&= -te^t.
\end{aligned}$$

Thus, the particular solution of (3.103) is $y_p(t) = -te^t$.

Example 3.16

Find a particular solution to the differential equation

$$y''(t) - 6y'(t) + 9y(t) = 5e^{3t}. \tag{3.104}$$

Solution

It is obvious that $\theta = 3$ is repeated root of $P(D) = D^2 - 6D + 9$. Thus, applying formula (3.100), we have

$$y_p(t) = \frac{5t^2 e^{\theta t}}{P''(\theta)}$$

$$= \frac{5t^2 e^{3t}}{P''(3)}$$

$$= \frac{5t^2 e^{3t}}{2}.$$

Consequently, the particular solution of (3.104) is

$$y_p(t) = \frac{5}{2} t^2 e^{3t}.$$

In the next theorem, we describe the superposition principle which is a powerful tool of finding particular solutions to a linear second order differential equations.

Theorem 3.3.6 (Superposition principle)

Let $y_1(t)$ be a solution to the differential equation

$$ay''(t) + by'(t) + cy(t) = f_1(t),$$

and $y_2(t)$ be a solution to

$$ay''(t) + by'(t) + cy(t) = f_2(t).$$

Then, for any constants k_1 and k_2, the function $k_1 y_1(t) + k_2 y_2(t)$ is a solution to the differential equation

$$ay''(t) + by'(t) + cy(t) = k_1 f_1(t) + k_2 f_2(t). \tag{3.105}$$

Proof

The proof is simple, we need just to verify that $Y(t) = k_1 y_1(t) + k_2 y_2(t)$ satisfies equation (3.105). Indeed, we have

$$aY''(t) + bY'(t) + cY(t) = k_1(ay_1'' + by_1' + cy_1) + k_2(ay_2'' + by_2' + cy_2)$$

$$= k_1 f_1(t) + k_2 f_2(t),$$

since $y_1(t)$ and $y_2(t)$ are solutions to the first and the second equation in Theorem 3.3.6, respectively. This finishes the proof of Theorem 3.3.6.

Example 3.17

Find a particular solution of the differential equation

$$y''(t) + 2y'(t) + 2y(t) = 5\sin t + 5\cos t. \tag{3.106}$$

Solution

According to the superposition principle (Theorem 3.3.6), the particular solution $Y_p(t)$ of (3.106) is defined as

$$Y_p(t) = y_p(t) + \hat{y}_p(t)$$

such that $y_p(t)$ is the particular solution of the equation

$$y''(t) + 2y'(t) + 2y(t) = 5\sin t$$

and $\hat{y}_p(t)$ is the particular solution of the equation

$$y''(t) + 2y'(t) + 2y(t) = 5\cos t.$$

Now, we consider the differential equation

$$y''(t) + 2y'(t) + 2y(t) = 5e^{it}. \tag{3.107}$$

Thus, if we denote by $\tilde{y}_p(t)$ as the particular solution of (3.107), then $y_p(t) = \Im(\tilde{y}_p(t))$ and $\hat{y}_p(t) = \Re(\tilde{y}_p(t))$. Since $\theta = i$ is not a root of $P(D) = D^2 + 2D + 2$, then applying formula (3.91), we get

$$\tilde{y}_p(t) = \frac{5e^{it}}{P(i)}$$
$$= \frac{5e^{it}}{1 + 2i}$$
$$= (\cos t + 2\sin t) + i(\sin t - 2\cos t).$$

Therefore,

$$y_p(t) = \sin t - 2\cos t, \quad \text{and} \quad \hat{y}_p(t) = \cos t + 2\sin t.$$

Consequently,

$$Y_p(t) = 3\sin t - \cos t.$$

Method of Undetermined Coefficients Here, we introduce a method of finding particular solutions of (3.87) in the case where $f(t)$ has one of several special form. This method requires us to make an initial guess about the form of the particular solution, depending on the form of $f(t)$, but with the coefficients of the particular solution left unspecified. We then substitute this assumed expression of the particular solution into equation (3.87) and attempt to find the coefficients so as to satisfy that equation. We summarize this method in the following theorem.

Theorem 3.3.7 (Undetermined coefficients)
We consider the differential equation

$$ay''(t) + by'(t) + cy(t) = Ct^m e^{rt}, \tag{3.108}$$

where α is a real number and C is a constant. Then, the particular solution of (3.108) has the form

$$y_p(t) = t^s \left(A_m t^m + A_{m-1} t^{m-1} + \ldots + A_1 t + A_0 \right) e^{rt}, \tag{3.109}$$

where

- $s = 0$, if r is not a root of $P(D) = aD^2 + bD + c$;
- $s = 1$, if r is a simple root of $P(D)$;
- $s = 2$, if r is a repeated root of $P(D)$.

If the term in the right-hand side in (3.108) has the form

$$Ct^m e^{\alpha t} \cos \beta t, \quad \text{or} \quad Ct^m e^{\alpha t} \sin \beta t,$$

then the particular solution is given by

$$y_p(t) = t^s \left(A_m t^m + A_{m-1} t^{m-1} + \ldots + A_1 t + A_0 \right) e^{\alpha t} \cos \beta t$$
$$+ t^s \left(B_m t^m + B_{m-1} t^{m-1} + \ldots + B_1 t + B_0 \right) e^{\alpha t} \sin \beta t, \tag{3.110}$$

where

- $s = 0$, if $\alpha + i\beta$ is not a root of $P(D) = aD^2 + bD + c$;
- $s = 1$, if $\alpha + i\beta$ is a simple root of $P(D)$.

Proof
We need just to prove (3.109) since (3.110) can be easily derived from (3.109) by taking $r = \alpha + i\beta$ and applying the Euler formula (3.25).

Assume that the function $f(t)$ has the form $Ct^m e^{rt}$. In this case, we need to look for a solution $y_p(t)$ such that the functions ay_p'', by_p' and cy_p add up to give a polynomial of order m multiplied by e^{rt}. The obvious choice of $y_p(t)$ is a polynomial of order n multiplied by e^{rt}. That is

$$y_p(t) = (A_n t^n + A_{n-1} t^{n-1} + \ldots + A_1 t + A_0) e^{rt}$$

$$= \left(\sum_{k=0}^{n} A_k t^k \right) e^{rt}$$

$$= \left(A_n t^n + A_{n-1} t^{n-1} \right) e^{rt} + \left(\sum_{k=0}^{n-2} A_k t^k \right) e^{rt}$$

with the power n to be determined later. By taking the derivative with respect to t, it then follows that

$$y_p'(t) = e^{rt} \left(\sum_{k=1}^{n} k A_k t^{k-1} \right) + r e^{rt} \left(\sum_{k=0}^{n} A_k t^k \right)$$

$$= e^{rt} \left(\sum_{k=0}^{n-1} (k+1) A_{k+1} t^k \right) + r e^{rt} \left(\sum_{k=0}^{n} A_k t^k \right)$$

$$= r e^{rt} (A_n t^n) + e^{rt} \sum_{k=0}^{n-1} \left((k+1) A_{k+1} + r A_k \right) t^k$$

$$= r e^{rt} (A_n t^n) + e^{rt} \left(n A_n + r A_{n-1} \right) t^{n-1} + e^{rt} \sum_{k=0}^{n-2} \left((k+1) A_{k+1} + r A_k \right) t^k.$$

Similarly, differentiating this last formula once again with respect to t, we get

$$y_p''(t) = re^{rt}(nA_nt^{n-1}) + r^2e^{rt}(A_nt^n)$$

$$+ e^{rt}\sum_{k=1}^{n-1}\left(k(k+1)A_{k+1} + rkA_k\right)t^{k-1} + re^{rt}\sum_{k=0}^{n-1}\left((k+1)A_{k+1} + rA_k\right)t^k$$

This gives by shifting the index,

$$y_p''(t) = e^{rt}\left(rnA_nt^{n-1} + r^2A_nt^n\right)$$

$$+ e^{rt}\sum_{k=0}^{n-2}\left((k+1)(k+2)A_{k+2} + r(k+1)A_{k+1}\right)t^k$$

$$+ re^{rt}\sum_{k=0}^{n-1}((k+1)A_{k+1} + rA_k)t^k$$

$$= e^{rt}\left(rnA_nt^{n-1} + r^2A_nt^n\right) + e^{rt}\left(rnA_n + r^2A_{n-1}\right)t^{n-1}$$

$$+ e^{rt}\sum_{k=0}^{n-2}\left((k+1)(k+2)A_{k+2} + 2r(k+1)A_{k+1} + r^2A_k\right)t^k$$

Arranging the above formula, we obtain

$$y_p''(t) = e^{rt}\left(r^2A_nt^n + (2rnA_n + r^2A_{n-1})t^{n-1}\right)$$

$$+ e^{rt}\sum_{k=0}^{n-2}\left((k+1)(k+2)A_{k+2} + 2r(k+1)A_{k+1} + r^2A_k\right)t^k.$$

Inserting the above formulas into equation (3.108), we find

$$ay_p''(t) + by_p'(t) + cy_p(t) = Ct^me^{rt}.$$

That is

$$ae^{rt}\left(r^2A_nt^n + (2rnA_n + r^2A_{n-1})t^{n-1}\right)$$

$$+ be^{rt}\left(rA_nt^n + (nA_n + rA_{n-1})t^{n-1}\right)$$

$$+ c(A_nt^n + A_{n-1}t^{n-1})e^{rt} + e^{rt}\sum_{k=0}^{n-2}B_kt^k = Ct^me^{rt} \qquad (3.111)$$

with

$$B_k = (ar^2 + br + c)A_k + (2ar + b)(k+1)A_{k+1} + a(k+1)(k+2)A_{k+2}.$$

We may simplify (3.111) and write it as

$$(ar^2 + br + c)A_nt^n + (ar^2 + br + c)A_{n-1}t^{n-1} + (2ar + b)nA_nt^{n-1} + \sum_{k=0}^{n-2}B_kt^k = Ct^m. \qquad (3.112)$$

Now, we have the following cases:

- **Case 1.** If r is not a root of $P(D)$, then $ar^2 + br + c \neq 0$ and therefore the leading term on the left-hand side of (3.112) is $(ar^2 + br + c)A_nt^n$ and to match the right-hand side, we need to take $n = m$. Therefore,

$$y_p(t) = \left(A_mt^m + A_{m-1}t^{m-1} + \ldots + A_1t + A_0\right)e^{rt}.$$

▬ **Case 2.** If r is a simple root of $P(D)$, then $ar^2 + br + c = 0$ and $2ar + b \neq 0$, then in this case the leading therm will be $(2ar + b)nA_nt^{n-1}$ and to match with Ct^m, we need to take $n = m + 1$. Thus,

$$y_p(t) = \left(A_{m+1}t^{m+1} + A_mt^m + \ldots + A_1t + A_0\right)e^{rt}.$$

However, the term A_0e^{rt} can be dropped since it solves the associated homogeneous equation. Consequently, the particular solution is given by (for simplicity, we renumber the coefficients)

$$y_p(t) = t\left(A_mt^m + A_{m-1}t^m + \ldots + A_0\right)e^{rt}.$$

▬ **Case 3.** If r is a repeated root of $P(D)$, then $ar^2 + br + c = 0$ and $2ar + b = 0$, and therefore the leading term in this case will be $B_{n-2}t^{n-2}$ and we should choose $n = m + 2$ to mach the right-hand side in (3.112). Consequently, the particular solution is

$$y_p(t) = \left(A_{m+2}t^{m+2} + A_{m+1}t^{m+1} + \ldots + A_1t + A_0\right)e^{rt}.$$

Once again we drop the term A_0e^{rt} since it is a solution of the associated homogeneous equation and the term A_1te^{rt} can be also dropped since $2ar + b = 0$. Thus, by renumbering the coefficients, we may write $y_p(t)$ as

$$y_p(t) = t^2\left(A_mt^m + A_{m-1}t^{m-1} + \ldots + A_1t + A_0\right)e^{rt}.$$

This completes the proof of Theorem 3.3.7.

Example 3.18

Find a particular solution to the differential equation

$$y''(t) + 3y'(t) + 2y(t) = 3t. \tag{3.113}$$

Solution

In order to find the particular solution of (3.113), we apply formula (3.109) with $\alpha = 0$, $C = 3$ and $m = 1$, and since $\alpha = 0$ is not a root to $P(D) = D^2 + 3D + 2$, then we obtain

$$y_p(t) = A_1t + A_0.$$

To find the constants A_1 and A_0 in the above equation, we use the fact that $y_p(t)$ satisfies equation (3.113), that is

$$y_p''(t) + 3y_p'(t) + 2y_p(t) = 3t.$$

This leads to

$$3A_1 + 2(A_1t + A_0) = 3t$$

which gives, by equating coefficients of like powers of t,

$$\begin{cases} 2A_1 = 3, \\ 3A_1 + 2A_0 = 0. \end{cases}$$

Solving the above algebraic system, we find $A_1 = 3/2$ and $A_0 = -9/4$. Consequently, the particular solution of (3.113) is given by

$$y_p(t) = \frac{3}{2}t - \frac{9}{4}.$$

Example 3.19

Find a particular solution to the differential equation

$$y''(t) - 3y'(t) + 2y(t) = e^t \sin t. \tag{3.114}$$

Solution

Applying formula (3.110) for $\alpha = 1$, $C = 1$, $\beta = 1$ and $m = 0$, we have the particular solution $y_p(t)$ of (3.114) defined as

$$y_p(t) = e^t (A_0 \cos t + B_0 \sin t), \tag{3.115}$$

since $(1 + i)$ is not a root of $P(D) = D^2 - 3D + 2$. To find the constants A_0 and B_0 in (3.115), we use the fact that $y_p(t)$ is a solution to (3.114), then we have

$$y_p''(t) - 3y_p'(t) + 2y_p(t) = e^t \sin t. \tag{3.116}$$

On the other hand, we have

$$y_p'(t) = \left((A_0 + B_0) \cos t + (B_0 - A_0) \sin t \right) e^t$$

and

$$y_p''(t) = \left(2B_0 \cos t - 2A_0 \sin t \right) e^t.$$

Inserting the above equalities into (3.116), we obtain

$$\left(2B_0 \cos t - 2A_0 \sin t \right) e^t - 3 \left((A_0 + B_0) \cos t + (B_0 - A_0) \sin t \right) e^t$$
$$+ 2(A_0 \cos t + B_0 \sin t) e^t = e^t \sin t.$$

This yields

$$(-A_0 - B_0) \cos t + (A_0 - 2B_0) \sin t = \sin t.$$

Consequently, it is enough to choose the constants A_0 and B_0 as

$$\begin{cases} A_0 + B_0 = 0, \\ A_0 - B_0 = 1. \end{cases}$$

The above algebraic system leads to $A_0 = 1/2$ and $B_0 = -1/2$. Finally, the particular solution of (3.114) is given by

$$y_p(t) = \frac{1}{2}(\cos t - \sin t) e^t.$$

3.3.2 Resonance

The phenomenon of *resonance* is well known to anyone who has pushed a child on a swing. If we choose the right time for pushing and the right frequency, the swing will eventually go very high, even with a slight push. For example resonance occurs when a structural system is forced at a frequency equals to the natural frequency of the system (the frequency at which the system naturally likes to vibrate). The effect of resonance can be destructive. It has been known that the mechanical resonance was responsible for the collapse of many bridges and buildings. For instance, the collapses of the Tacoma Bridge in 1940 in Washington and the Broughton Suspension Bridge in 1831 near Manchester were caused by the effect of resonance.

In order to explain this physical phenomenon, we consider the following second order differential equation

$$y''(t) + \omega_0^2 y(t) = \cos \omega_1 t. \tag{3.117}$$

This is the case of the harmonic oscillator studied in ▶ Sect. 3.1.3 with the damping coefficient $b = 0$ and we add the forcing (or driving) term $f(t) = \cos \omega_1 t$.

Let us assume that the *driving frequency* ω_1 is different from the *natural frequency* (usually called the *eigenfrequency* $\omega_0 = (k/m)^{1/2}$ of the undamped oscillator) i.e. $\omega_1 \neq \omega_0$. The particular solution $y_p(t)$ of (3.117) is the real part of the solution $\tilde{y}(t)$ of the equation

$$P(D)[\tilde{y}(t)] = e^{i\omega_1 t}, \qquad \text{with} \qquad P(D) = D^2 + \omega_0^2. \tag{3.118}$$

Since $\omega_1 \neq \omega_0$, then applying formula (3.91), we get

$$\begin{aligned}
\tilde{y}_p(t) &= \frac{e^{i\omega_1 t}}{P(i\omega_1)} \\
&= \frac{e^{i\omega_1 t}}{\omega_0^2 - \omega_1^2}.
\end{aligned}$$

Consequently, the particular solution of (3.117) (it is also called the *steady state* solution) is given by

$$\begin{aligned}
y_p(t) &= \Re\{\tilde{y}_p(t)\} \\
&= \frac{\cos \omega_1 t}{\omega_0^2 - \omega_1^2}. \tag{3.119}
\end{aligned}$$

Formula (3.119) shows that the response $y_p(t)$ oscillates with the same frequency of the forcing term ω_1 and with an amplitude that depends on the eigenfrequency ω_0 of the oscillator. As it is shown in (3.119) a remarkable thing happens when $\omega_1 \to \omega_0$, then the amplitude

$$A = \frac{1}{\omega_0^2 - \omega_1^2}$$

becomes very large and tends to infinity. In other words, the response of the oscillator grows without bounds and resonance occurs.

Now, if $\omega_1 = \omega_0$, then equation (3.118) can be written as

$$P(D)[\tilde{y}(t)] = e^{i\omega_0 t}, \qquad \text{with} \qquad P(D) = D^2 + \omega_0^2. \tag{3.120}$$

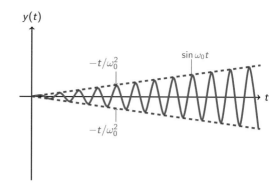

In this case ω_0 is a simple root of $P(D)$. Thus, applying formula (3.99), we obtain

$$\tilde{y}_p(t) = \frac{t e^{i\omega_0 t}}{P'(i\omega_0)}$$
$$= \frac{t e^{i\omega_0 t}}{2i\omega_0}.$$

Therefore, by taking the real part of the above solution, we find that the particular solution of (3.117) for $\omega_1 = \omega_0$ has the form

$$y_p(t) = \frac{t \sin \omega_0 t}{2\omega_0} \tag{3.121}$$

and it is illustrated in ■ Fig. 3.6.

Now, we want to show that the study state solution

$$y_p(t) = \frac{t \sin \omega_0 t}{2\omega_0}$$

is exactly the limit when ω_1 goes to ω_0 of the general solution of (3.117) satisfying the initial conditions $y(0) = 0$ and $y'(0) = 0$. This solution is defined by

$$y(t) = \frac{\cos \omega_1 t}{\omega_0^2 - \omega_1^2} - \frac{\cos \omega_0 t}{\omega_0^2 - \omega_1^2} \tag{3.122}$$

Now, we compute

$$\lim_{\omega_1 \to \omega_0} y(t) = \lim_{\omega_1 \to \omega_0} \frac{\cos \omega_1 t - \cos \omega_0 t}{\omega_0^2 - \omega_1^2}$$
$$= \frac{t \sin \omega_0 t}{2\omega_0},$$

where we have used L'Hôpital's rule. This gives our desired result.

3.3.3 Method of Variation of Parameters

We have introduced in ▶ Sect. 3.3.1 a useful method to find a particular solution to the second order linear inhomogeneous equation

$$a(t)y''(t) + b(t)y'(t) + c(t)y(t) = f(t). \tag{3.123}$$

Unfortunately, the method of undetermined coefficients has two severe limitations: it can be used only when the coefficients $a(t)$, $b(t)$ and $c(t)$ are constants and even then, it is effective in only few special cases when the input unction $f(t)$ is an exponential, sine or cosine, a polynomial, or some combination of such functions. Within these limitations, however, that procedure is usually the easiest to apply.

We now develop a more powerful method that always works regardless of the nature of $a(t)$, $b(t)$, $c(t)$ and $f(t)$ provided only that the general solution of the corresponding homogenous equation

$$a(t)y''(t) + b(t)y'(t) + c(t)y(t) = 0, \tag{3.124}$$

is already known.

The method of *variation of parameters* is similar to the one discussed in ▶ Sect. 3.2.1. Assume that we know two linearly independent solutions $y_1(t)$ and $y_2(t)$ to the homogeneous equation (3.124). Thus, the general solution of (3.124) is written in the form

$$y(t) = c_1 y_1(t) + c_2 y_2(t). \tag{3.125}$$

The basic idea of the method of variation of parameters is to replace the constants c_1 and c_2 in (3.125) by unknown functions $v_1(t)$ and $v_2(t)$, and attempt to determine $v_1(t)$ and $v_2(t)$ in such a manner that

$$y_p(t) = v_1(t)y_1(t) + v_2(t)y_2(t) \tag{3.126}$$

will be a solution to (3.123). To find these two functions $v_1(t)$ and $v_2(t)$, it is necessary to have two equations relating these functions. Naturally, one of these equations will come from the fact that (3.126) is a solution of (3.123). To accomplish this, we compute the first derivative of (3.126), we obtain (we drop the t for simplicity)

$$y_p' = (v_1'y_1 + v_2'y_2) + (v_1y_1' + v_2y_2'). \tag{3.127}$$

In order to simplify the computations and to avoid the second order derivatives of the unknown functions v_1 and v_2, we require the first expression between parenthesis in (3.127) to vanish. That is

$$v_1'y_1 + v_2'y_2 = 0. \tag{3.128}$$

Consequently, the equation (3.127) takes the form

$$y'_p = v_1 y'_1 + v_2 y'_2.$$

(3.129)

The second derivatives of $y_p(t)$ becomes

$$y''_p = v'_1 y'_1 + v_1 y''_1 + v'_2 y'_2 + v_2 y''_2.$$

(3.130)

Plugging (3.126), (3.127) and (3.130) into (3.123) and keeping in mind that y_1 and y_2 are solutions to the homogeneous equation (3.124), we get

$$v_1(a y''_1 + b y'_1 + c y_1) + v_2(a y''_2 + b y'_2 + c y_2) + a(v'_1 y'_1 + v'_2 y'_2) = f.$$

This leads to the equation

$$v'_1 y'_1 + v'_2 y'_2 = \frac{f}{a}.$$

(3.131)

Collecting (3.128) and (3.131) together, we have the algebraic system of two equations with the two unknowns v'_1 and v'_2:

$$\begin{cases} y_1 v'_1 + y_2 v'_2 = 0, \\ y'_1 v'_1 + y'_2 v'_2 = \dfrac{f}{a}. \end{cases}$$

(3.132)

System (3.132) can be solved as

$$v'_1 = -\frac{1}{a} \frac{y_2 f(t)}{W[y_1, y_2]}, \qquad \text{and} \qquad v'_2 = \frac{1}{a} \frac{y_1 f(t)}{W[y_1, y_2]},$$

(3.133)

where $W[y_1, y_2]$ is the Wronskian defined in (3.3), and since y_1 and y_2 are linearly independent, the $W[y_1, y_2] \neq 0$. All that remains is to integrate the formulas in (3.133) to find v_1 and v_2 as

$$v_1 = -\frac{1}{a} \int \frac{y_2 f(t)}{W[y_1, y_2]} dt, \qquad \text{and} \qquad v_2 = \frac{1}{a} \int \frac{y_1 f(t)}{W[y_1, y_2]} dt.$$

(3.134)

Consequently, by inserting the above formulas into (3.126), we obtain

$$y_p = \frac{1}{a} \left[-y_1 \int \frac{y_2 f(t)}{W[y_1, y_2]} dt + y_2 \int \frac{y_1 f(t)}{W[y_1, y_2]} dt \right]$$

as the particular solution of (3.123).

We summarize the above discussion in the following theorem.

Theorem 3.3.8
Let $y_1(t)$ and $y_2(t)$ be two independent solutions of the homogeneous equation

$$a(t)y''(t) + b(t)y'(t) + c(t)y(t) = 0.$$

Then a particular solution of the non-homogeneous equation

$$a(t)y''(t) + b(t)y'(t) + c(t)y(t) = f(t),$$

is given by

$$y_p(t) = \frac{1}{a}\left[-y_1(t)\int \frac{y_2(t)f(t)}{W[y_1, y_2]}dt + y_2(t)\int \frac{y_1(t)f(t)}{W[y_1, y_2]}dt\right], \qquad (3.135)$$

where $W[y_1, y_2]$ is the Wronskian of the two independent solutions $y_1(t)$ and $y_2(t)$.

ℹ Remark 3.3.9 The method of variation of parameters has some disadvantages of its own. In particular, the integrals in (3.135) may be difficult or impossible to handle. In addition, it is necessary to know the general solution of the corresponding homogeneous equation before we can even start the process of finding a particular solution.

Example 3.20
Find a particular solution of the differential equation

$$y''(t) + y(t) = \frac{1}{\sin t}, \qquad (3.136)$$

in the interval $(0, \pi/2)$.

Solution
To find a particular solution of (3.136), we need first to find two independent solutions $y_1(t)$ and $y_2(t)$ to the homogeneous equation

$$y''(t) + y(t) = 0. \qquad (3.137)$$

The characteristic equation associated to (3.137) is

$$r^2 + 1 = 0,$$

which has the two complex roots $r_1 = i$ and $r_2 = -i$. Thus, the general solution of (3.139) has the form

$$y(t) = c_1 \cos t + c_2 \sin t.$$

It is obvious that $y_1(t) = \cos t$ and $y_2(t) = \sin t$ are two independent solutions of (3.139) since

$$W[y_1, y_2] = y_1(t)y_2'(t) - y_2(t)y_1'(t)$$
$$= \cos^2 t + \sin^2 t = 1.$$

Applying formulas (3.134), we have

$$v_1(t) = -\int \sin t \frac{1}{\sin t} dt = -t.$$

and

$$v_2(t) = \int \frac{\cos t}{\sin t} dt = \ln(\sin t).$$

Consequently, the particular solution of (3.136) is

$$y_p(t) = v_1(t)y_1(t) + v_2(t)y_2(t)$$
$$= -t \cos t + \ln(\sin t) \cdot (\sin t).$$

Example 3.21

Find a particular solution of the differential equation

$$y''(t) + y(t) = \tan t \tag{3.138}$$

in the interval $(-\pi/2, \pi/2)$.

Solution

To find a particular solution of (3.138), we need first to find two independent solutions $y_1(t)$ and $y_2(t)$ to the homogeneous equation

$$y''(t) + y(t) = 0. \tag{3.139}$$

As we have seen in Example 3.20, $y_1(t) = \cos t$ and $y_2(t) = \sin t$ are two independent solutions of (3.139) with

$$W[y_1, y_2] = 1.$$

We look for a particular solution of the form

$$y_p(t) = v_1(t) \cos t + v_2(t) \sin t,$$

where

$$\begin{aligned} v_1(t) = -\int \sin t \tan t \, dt &= -\int \frac{\sin^2 t}{\cos t} dt \\ &= \int \frac{\cos^2 t - 1}{\cos t} dt \\ &= \int (\cos t - \sec t) dt \\ &= \sin t + \ln|\sec t + \tan t| \end{aligned}$$

and

$$v_2(t) = \int \cos t \tan t \, dt = \int \sin t \, dt = -\cos t.$$

Consequently, the particular solution of (3.138) is

$$\begin{aligned} y_p(t) &= (\cos t) \cdot (\sin t + \ln|\sec t + \tan t|) - (\sin t) \cdot (\cos t) \\ &= (\cos t) \cdot (\ln|\sec t + \tan t|). \end{aligned}$$

3.3.4 Exercises

Exercise 3.1

Solve the differential equation

$$y''(t) - 3y'(t) + 2y(t) = 2t^3 - 7t^2 + 2t - 1. \tag{3.140}$$

Solution

Equation (3.140) is a second order differential equation with constant coefficients. Thus, using Theorem 3.3.1, the solution of (3.140) can be written as

$$y(t) = y_h(t) + y_p(t),$$

where $y_h(t)$ is the general solution of the homogeneous equation

$$y''(t) - 3y'(t) + 2y(t) = 0 \tag{3.141}$$

and $y_p(t)$ is a particular solution of (3.140). To find $y_h(t)$, we look first to the solutions of the characteristic equation associated to (3.141):

$$r^2 - 3r + 2 = 0.$$

This last equation has two real distinct roots $r_1 = 1$ and $r_2 = 2$. Thus, the general solution of (3.141) is given by

$$y_h(t) = c_1 e^t + c_2 e^{2t}, \tag{3.142}$$

where c_1 and c_2 are two positive constants.

Next, to find the particular solution $y_p(t)$, we apply the method of undetermined coefficients described in Theorem 3.3.7 and the superposition principle in Theorem 3.3.6, to deduce that the particular solution $y_p(t)$ has the form

$$y_p(t) = A_3 t^3 + A_2 t^2 + A_1 t + A_0,$$

where A_i, $i = 0, 1, 2, 3$ are constants that will be determined later.

The first and the second derivatives of $y_p(t)$ gives

$$y_p'(t) = 3A_3 t^2 + 2A_2 t + A_1, \quad \text{and} \quad y_p''(t) = 6A_3 t + 2A_2.$$

Plugging the above formulas into (3.140), we obtain

$$2A_3 t^3 + (2A_2 - 9A_3)t^2 + (2A_1 - 6A_2 + 6A_3)t + (2A_0 - 3A_1 + 2A_2) = 2t^3 - 7t^2 + 2t - 1$$

Consequently, equating the coefficients of like powers of t, we deduce from above that

$$\begin{cases} 2A_3 = 2, \\ 2A_2 - 9A_3 = -7, \\ 2A_1 - 6A_2 + 6A_3 = 2, \\ 2A_0 - 3A_1 + 2A_2 = -1. \end{cases}$$

The solution of the above system gives $A_1 = A_2 = A_3 = 1$ and $A_0 = 0$. Therefore,

$$y_p(t) = t^3 + t^2 + 1 \tag{3.143}$$

is a particular solution of (3.140). Collecting (3.142) and (3.143), we deduce that the solution of (3.140) is

$$y(t) = c_1 e^t + c_2 e^{2t} + t^3 + t^2 + 1.$$

Exercise 3.2

Find the general solution of the differential equation

$$y''(t) - 4y'(t) + 13y(t) = 10\cos 2t + 25\sin 2t. \tag{3.144}$$

Solution

As we have seen before, the general solution of (3.144) is

$$y(t) = y_h(t) + y_p(t),$$

where $y_h(t)$ is the general solution of the homogeneous equation

$$y''(t) - 4y'(t) + 13y(t) = 0 \tag{3.145}$$

and $y_p(t)$ is a particular solution of (3.144).

The characteristic equation associated to (3.145) is

$$r^2 - 4r + 13 = 0,$$

and has two complex roots $r_1 = 2 + 3i$ and $r_2 = 2 - 3i$. Thus, $y_h(t)$ is given by

$$y_h(t) = e^{2t}(c_1 \cos 3t + c_2 \sin 3t), \tag{3.146}$$

where c_1 and c_2 are two constants.

To find a particular solution of (3.144), we need first to find a particular solution $y_1(t)$ of

$$y''(t) - 4y'(t) + 13y(t) = \cos 2t \tag{3.147}$$

and a particular solution $y_2(t)$ of

$$y''(t) - 4y'(t) + 13y(t) = \sin 2t. \tag{3.148}$$

Then, the superposition principle leads to $y_p(t) = 10y_1(t) + 25y_2(t)$.

To find $y_1(t)$ and $y_2(t)$, we apply Theorem 3.3.3, we get

$$y_1(t) = \Re(\tilde{y}_p(t)), \quad \text{and} \quad y_2(t) = \Im(\tilde{y}_p(t))$$

where

$$\tilde{y}_p(t) = \frac{e^{\theta t}}{P(\theta)}, \quad \theta = 2i.$$

and $P(\theta) = \theta^2 - 4\theta + 13$. Thus,

$$\tilde{y}_p(t) = \frac{e^{2it}}{9 - 8i} = \left(\frac{9}{145}\cos 2t - \frac{8}{145}\sin 2t\right) + i\left(\frac{9}{145}\sin 2t + \frac{8}{145}\cos 2t\right).$$

Consequently,

$$y_1(t) = \frac{9}{145}\cos 2t - \frac{8}{145}\sin 2t, \quad \text{and} \quad y_2(t) = \frac{9}{145}\sin 2t + \frac{8}{145}\cos 2t$$

Thus,

$$y_p(t) = 10y_1(t) + 25y_2(t)$$
$$= 2\cos 2t + \sin 2t. \tag{3.149}$$

Finally, collecting (3.146) and (3.149), then the solution of (3.144) is

$$y(t) = e^{2t}(c_1 \cos 3t + c_2 \sin 3t) + 2\cos 2t + \sin 2t.$$

Exercise 3.3

Use the method of variation of parameters to find the general solution of the differential equation

$$y''(t) - y(t) = \frac{1}{\cosh^3 t}, \qquad (3.150)$$

where $\cosh t = \frac{1}{2}(e^t + e^{-t})$.

Solution

The solution of (3.150) is

$$y(t) = y_h(t) + y_p(t),$$

where $y_h(t)$ is the general solution of the homogeneous equation

$$y''(t) - y(t) = 0 \qquad (3.151)$$

and $y_p(t)$ is a particular solution of (3.150).

Let us first find the solution $y_h(t)$ of equation (3.151). Indeed, the characteristic equation associated to (3.151) is $r^2 - 1 = 0$ and its roots are $r_1 = 1$ and $r_2 = -1$. Since the roots are real and distinct, then as we have seen in ▶ Sect. 3.1.2,

$$y_h(t) = c_1 e^t + c_2 e^{-t}. \qquad (3.152)$$

Since the right-hand side of (3.150) is a term involving $\cosh t$, then, it is more convenient to write $y_h(t)$ in (3.152) as

$$\begin{aligned} y_h(t) &= c_1 e^t + c_2 e^{-t} \\ &= (c_1 + c_2)\left(\frac{e^t + e^{-t}}{2}\right) + (c_2 - c_1)\left(\frac{e^t - e^{-t}}{2}\right) \\ &= d_1 \cosh t + d_2 \sinh t \end{aligned} \qquad (3.153)$$

where $d_1 = c_1 + c_2$ and $d_2 = c_2 - c_1$.

Next, and in order to find $y_p(t)$, we use the method of variation of parameters in ▶ Sect. 3.3.3 and search for $y_p(t)$ in the form

$$y_p(t) = v_1(t)\cosh t + v_2(t)\sinh t. \qquad (3.154)$$

We put $y_1(t) = \cosh t$ and $y_2(t) = \sinh t$ and apply (3.134), we obtain

$$v_1 = -\int \frac{y_2}{W[y_1, y_2]} \frac{1}{\cosh^3 t} dt, \quad \text{and} \quad v_2 = \int \frac{y_1}{W[y_1, y_2]} \frac{1}{\cosh^3 t} dt.$$

We have

$$\begin{aligned} W[y_1, y_2] &= y_1(t)y_2'(t) - y_2(t)y_1'(t) \\ &= \cosh^2 t - \sinh^2 t = 1. \end{aligned}$$

Thus, plugging this into the above formula of v_1 and v_2, we get

$$v_1 = -\int \frac{\sinh t}{\cosh^3 t} dt, \quad \text{and} \quad v_2 = \int \frac{1}{\cosh^2 t} dt.$$

Now, to find v_1, we use the change of variable

$$u = \cosh t.$$

This gives

$$du = \sinh t\, dt.$$

Then, v_1 becomes

$$v_1 = -\int \frac{du}{u^3} = \frac{1}{2u^2}$$

$$= \frac{1}{2\cosh^2 t}.$$

On the other hand

$$v_2 = \int \frac{1}{\cosh^2 t}\, dt = \tanh t.$$

Inserting v_1 and v_2 into (3.154), we find

$$y_p(t) = \frac{1}{2\cosh t} + (\tanh t)\cdot(\sinh t). \tag{3.155}$$

Finally, from (3.153) and (3.155), we deduce that the general solution of (3.150) is

$$y(t) = d_1 \cosh t + d_2 \sinh t + \frac{1}{2\cosh t} + (\tanh t)\cdot(\sinh t).$$

Exercise 3.4
Find a general solution to the differential equation

$$y''(t) - 4y'(t) + 4y(t) = e^{2t}. \tag{3.156}$$

Solution
As we have seen before, we need first to find a solution $y_h(t)$ to the homogeneous equation

$$y''(t) - 4y'(t) + 4y(t) = 0. \tag{3.157}$$

After that, we have to search for a particular solution $y_p(t)$ of (3.156). Then, the general solution $y(t)$ of (3.156) is

$$y(t) = y_h(t) + y_p(t).$$

The characteristic equation associated to (3.157) is $r^2 - 4r + 4 = 0$ and has $r = 2$ as a repeated root. Using the method described in ▶ Sect. 3.1.2, we write $y_h(t)$ as

$$y_h(t) = (c_1 + c_2 t)e^{2t}. \tag{3.158}$$

To find $y_p(t)$ and since $\theta = 2$ is a repeated root of the polynomial $P(D) = D^2 - 4D + 4$, then we use the formula (3.100) to get

$$y_p(t) = \frac{t^2 e^{\theta t}}{P''(\theta)}$$

$$= \frac{t^2}{2} e^{2t}.$$

Consequently, the general solution of (3.156) is

$$y(t) = \left(c_1 + c_2 t + \frac{t^2}{2}\right)e^{2t},$$

where c_1 and c_2 are two constants.

Exercise 3.5

Find a general solution to the differential equation

$$y''(t) - 2y'(t) + 5y(t) = e^t \sin 2t + \cos t. \tag{3.159}$$

Solution

As in the above Exercises, we need first to find a solution $y_h(t)$ of the homogeneous equation

$$y''(t) - 2y'(t) + 5y(t) = 0. \tag{3.160}$$

Then, we need to find a particular solution $y_p(t)$ to (3.159). After that, we add the two solutions to obtain the general solution.

The characteristic equation associated to (3.160) is $r^2 - 2r + 5 = 0$ and has two complex roots: $r_1 = 1 + 2i$ and $r_2 = 1 - 2i$. Thus, $y_h(t)$ is given by

$$y_h(t) = e^t(c_1 \cos 2t + c_2 \sin 2t), \tag{3.161}$$

where c_1 and c_2 are two constants.

To find the particular solution $y_p(t)$, we need to use the superposition principle (Theorem 3.3.6). So, we need first to find a particular solution $y_1(t)$ of the differential equation

$$y''(t) - 2y'(t) + 5y(t) = e^t \sin 2t \tag{3.162}$$

and a particular solution $y_2(t)$ of the differential equation

$$y''(t) - 2y'(t) + 5y(t) = \cos t. \tag{3.163}$$

Then, by using Theorem 3.3.6, we get $y_p(t) = y_1(t) + y_2(t)$. Indeed, $y_1(t) = \Im(\tilde{y}_1(t))$, where $\tilde{y}_1(t)$ is the solution of the equation

$$y''(t) - 2y'(t) + 5y(t) = e^{(1+2i)t}. \tag{3.164}$$

Since $\theta = 1 + 2i$ is a root of $P(D) = D^2 - 2D + 5$, then using formula (3.99), we get

$$\tilde{y}_1(t) = \frac{te^{\theta t}}{P'(\theta)}$$
$$= -\frac{it}{4}e^t(\cos 2t + i \sin 2t).$$

Therefore,

$$y_1(t) = \Im(\tilde{y}_1(t)) = -\frac{t}{4}e^t \cos 2t.$$

On the other hand, $y_2(t) = \Re(\tilde{y}_2(t))$, where $\tilde{y}_2(t)$ is the solution of the equation

$$y''(t) - 2y'(t) + 5y(t) = e^{it}.$$

Since $\theta = i$ is not a root to $P(D)$, then we have, by using (3.91)

$$\tilde{y}_2(t) = \frac{e^{\theta t}}{P(\theta)}$$

$$= \frac{e^{it}}{4 - 2i}.$$

Thus,

$$y_2(t) = \Re(\tilde{y}_2(t)) = \frac{1}{5}\cos t - \frac{1}{10}\sin t.$$

Consequently,

$$y_p(t) = -\frac{t}{4}e^t \cos 2t + \frac{1}{5}\cos t - \frac{1}{10}\sin t. \tag{3.165}$$

Finally, collecting (3.161) and (3.165), we deduce that the general solution of (3.159) is

$$y(t) = e^t(c_1 \cos 2t + c_2 \sin 2t) - \frac{t}{4}e^t \cos 2t + \frac{1}{5}\cos t - \frac{1}{10}\sin t.$$

Exercise 3.6

Find for $t > 0$ a general solution to the differential equation

$$y''(t) + 2y'(t) + y(t) = e^{-t}\ln t. \tag{3.166}$$

Solution

Let us first find the solution $y_h(t)$ of the homogeneous equation

$$y''(t) + 2y'(t) + y(t) = 0. \tag{3.167}$$

The characteristic equation associated to (3.167) is $r^2 + 2r + 1 = 0$ and has $r = -1$ as a repeated root. Therefore, the solution $y_h(t)$ can be written as

$$y_h(t) = (c_1 + c_2 t)e^{-t}, \tag{3.168}$$

where c_1 and c_2 are two constants.

To find the particular solution $y_p(t)$ of (3.166), we use the method of variation of parameters introduced in ▶ Sect. 3.3.3. Thus, we write $y_p(t)$ as

$$y_p(t) = v_1(t)e^{-t} + v_2(t)te^{-t},$$
$$= v_1(t)y_1(t) + v_2(t)y_2(t),$$

where $y_1(t) = e^{-t}$ and $y_2(t) = te^{-t}$ and v_1 and v_2 are as follows:

$$v_1(t) = -\int \frac{te^{-2t}\ln t}{W[y_1, y_2]}dt, \quad \text{and} \quad v_2(t) = \int \frac{e^{-2t}\ln t}{W[y_1, y_2]}dt,$$

where

$$W[y_1, y_2] = y_1(t)y_2'(t) - y_2(t)y_1'(t)$$
$$= e^{-2t}.$$

Consequently, $v_1(t)$ and $v_2(t)$ take the form

$$v_1(t) = -\int t \ln t \, dt, \qquad \text{and} \qquad v_2(t) = \int \ln t \, dt.$$

Using the integration by parts, we obtain

$$v_1(t) = -\frac{t^2}{2} \ln t + \frac{t^2}{4}, \qquad \text{and} \qquad v_2(t) = t \ln t - t.$$

Consequently, $y_p(t)$ takes the form

$$y_p(t) = \frac{1}{2} t^2 e^{-t} \ln t - \frac{3}{4} t^2 e^{-t}. \tag{3.169}$$

Finally, collecting (3.168) and (3.169), we deduce that the general solution of (3.166) has the form

$$y(t) = (c_1 + c_2 t) e^{-t} + \frac{1}{2} t^2 e^{-t} \ln t - \frac{3}{4} t^2 e^{-t}.$$

Laplace Transforms

Belkacem Said-Houari

B. Said-Houari, *Differential Equations: Methods and Applications*, Compact Textbooks in Mathematics, DOI 10.1007/978-3-319-25735-8_4, © Springer International Publishing Switzerland 2015

4.1 Introduction and Definition

One of the powerful tools and efficient method of solving certain types of differential equations is the *Laplace transform*, which transform the original differential equation into an algebraic expression which can then simply be transformed once again using the *inverse Laplace transform* into the solution of the original problem.

Before entering into the details, let us first make a connection between the Laplace transform and the power series. We consider a power series in the form

$$\sum_{n=0}^{\infty} a(n)x^n. \tag{4.1}$$

If the series in (4.1) converges, then we may write its sum as a function of x. That is

$$\sum_{n=0}^{\infty} a(n)x^n = A(x). \tag{4.2}$$

The function $A(x)$ in (4.2) depends on crucial way on the form of the coefficients $a(n)$. For instance if we take in (4.2) $a(n) = 1$ for all n, then we get the geometric series

$$\sum_{n=0}^{\infty} x^n$$

and its sum is

$$\sum_{n=0}^{\infty} x^n = A(x) = \frac{1}{1-x},$$

provided that $|x| < 1$. Otherwise the geometric series will not converge. Another important example is to consider $a(n) = 1/n!$. Then, in this case, we obtain

$$\sum_{n=0}^{\infty} \frac{1}{n!}x^n = A(x) = e^x,$$

which holds for all x in $\mathbb{R}$.

Now, the continuous analogs of power series is to consider instead of the discrete values of $n = 0, 1, \ldots$, the continuous variable t and $0 \le t < \infty$. Thus, by changing the sum in (4.1) to an integral, we get the following integral:

$$\int_0^\infty a(t)x^t dt. \tag{4.3}$$

In many situation the improper integral in (4.3) does not converge, especially when $x > 1$. So, it is more convenient to assume $0 < x < 1$ in (4.3). In this case the integral in (4.3) can be written as

$$\int_0^\infty a(t)e^{t \ln x} dt. \tag{4.4}$$

Since $0 < x < 1$, then $\ln x < 0$, Thus, we put $s = -\ln x$ and rewrite (4.4) in the form

$$\int_0^\infty a(t)e^{-st} dt. \tag{4.5}$$

If the integral in (4.5) converges, then we have

$$\int_0^\infty a(t)e^{-st} dt = A(x) = A(e^{-s}).$$

If we put $F(s) = A(e^{-s})$ and write $f(t)$ instead of $a(t)$, then the above integral leads to

$$\int_0^\infty f(t)e^{-st} dt = F(s).$$

The function $F(s)$ in the above formula is the *Laplace transform* of the function $f(t)$.

Definition 4.1.1

Let $f(t)$ be a real-valued function on $[0, \infty)$. The *Laplace transform* of $f(t)$ is the function $F(s)$ defined by the integral

$$F(s) = \int_0^\infty f(t)e^{-st} dt, \quad s > 0. \tag{4.6}$$

The domain of $F(s)$ is all the values of s for which the improper integral in (4.6) exists. The Laplace transform of f is denoted by both $F(s)$ and $\mathcal{L}\{f\}(s)$.

Thus, the Laplace transform $\mathcal{L}$ acts on any function f for which the integral in (4.6) exists and produce its Laplace transform $\mathcal{L}\{f\}(s) = F(s)$, a function of the parameter s. The improper integral in (4.6) is defined to be the following limit, and exists only when this limit exists:

$$\int_0^\infty f(t)e^{-st}\,dt = \lim_{N\to\infty}\int_0^N f(t)e^{-st}\,dt. \tag{4.7}$$

> **ℹ Remark 4.1.1** Similarly, we can define the Laplace transform of a complex-valued function $f(t)$ as

$$F(s) = \int_0^\infty f(t)e^{-st}\,dt, \quad \Re(s) > 0.$$

But, here we are interested only in the real-valued functions.

Example 4.1
Prove that the Laplace transform of $f(t) = 1$ is

$$F(s) = \frac{1}{s}, \quad s > 0.$$

Solution
Let us take $f(t) = 1$ for all $t \geq 0$. Then, we have

$$F(s) = \int_0^\infty e^{-st}\,dt = \lim_{N\to\infty}\int_0^N e^{-st}\,dt$$

$$= \lim_{N\to\infty}\left[\frac{-e^{-st}}{s}\right]_{t=0}^{t=N}$$

$$= \lim_{N\to\infty}\left[\frac{1}{s} - \frac{e^{-sN}}{s}\right]$$

$$= \frac{1}{s},$$

for $s > 0$. Since $\lim_{N\to\infty} e^{-sN} = 0$ for $s > 0$.

For $s \leq 0$, the Laplace transform of $f(t) = 1$ is not defined since the above integral diverges. Consequently, if $f(t) = 1$, for all $t \geq 0$, the

$$F(s) = \frac{1}{s}, \quad s > 0.$$

Example 4.2
Show that for any constant a, the Laplace transform of $f(t) = e^{at}$ is

$$F(s) = \frac{1}{s-a}, \quad s > a.$$

Solution

Using formula (4.7), we have

$$F(s) = \int_0^\infty e^{at} e^{-st}\, dt$$

$$= \lim_{N \to \infty} \int_0^N e^{-(s-a)t}\, dt$$

$$= \lim_{N \to \infty} \left[\frac{-e^{-(s-a)t}}{s-a} \right]_{t=0}^{t=N}$$

$$= \lim_{N \to \infty} \left[\frac{1}{s-a} - \frac{-e^{-(s-a)N}}{s-a} \right]$$

$$= \frac{1}{s-a},$$

for $s > a$. If $s \le a$, then the Laplace transform of $f(t) = e^{at}$ is not defined.

Example 4.3

Show the following

$$\mathcal{L}\{\sin bt\}(s) = \frac{b}{s^2 + b^2}, \qquad s > 0 \tag{4.8}$$

and

$$\mathcal{L}\{\cos bt\}(s) = \frac{s}{s^2 + b^2}, \qquad s > 0. \tag{4.9}$$

Solution

To show (4.8), we make use of (4.7) to get

$$\int_0^\infty e^{-st} \sin bt\, dt = \lim_{N \to \infty} \int_0^N e^{-st} \sin bt\, dt.$$

Using integration by parts, then the right-hand side in the above formula gives

$$\lim_{N \to \infty} \int_0^N e^{-st} \sin bt\, dt = \lim_{N \to \infty} \left[\frac{e^{-st}}{s^2 + b^2} (-s \sin bt - b \cos bt) \right]_{t=0}^{t=N}$$

$$= \lim_{N \to \infty} \left[\frac{b}{s^2 + b^2} - \frac{e^{-sN}}{s^2 + b^2} (s \sin bN + b \cos bN) \right]$$

$$= \frac{b}{s^2 + b^2},$$

since $s > 0$. This gives (4.8). We can easily show (4.9). We leave the details to the reader.

Example 4.4

Prove that for any positive integer n, we have

$$\mathcal{L}\{t^n\}(s) = \frac{n!}{s^{n+1}}, \qquad n = 0, 1, 2, \ldots \qquad s > 0. \tag{4.10}$$

Solution

Formula (4.10) can be proved by induction. Indeed, it is clear from Example 4.1 that (4.10) holds true for $n = 0$. Let us assume that (4.10) is true for $n - 1$ that is

$$\mathcal{L}\{t^{n-1}\}(s) = \frac{(n-1)!}{s^n}, \qquad s > 0 \tag{4.11}$$

and show that it remains also true for n. Using (4.7) and performing an integration by parts, we get

$$\mathcal{L}\{t^n\}(s) = \lim_{N \to \infty} \int_0^N t^n e^{-st} \, dt$$

$$= \lim_{N \to \infty} \left[-\frac{t^n e^{-st}}{s} \right]_{t=0}^{t=N} + \lim_{N \to \infty} \frac{n}{s} \int_0^N t^{n-1} e^{-st} \, dt.$$

It is clear that for $s > 0$, then $\lim_{N \to \infty} t^n e^{-st} = 0$. Thus, we get from above

$$\mathcal{L}\{t^n\}(s) = \frac{n}{s} \int_0^\infty t^{n-1} e^{-st} \, dt$$

$$= \frac{n}{s} \mathcal{L}\{t^{n-1}\}(s)$$

$$= \frac{n}{s} \frac{(n-1)!}{s^n}$$

$$= \frac{n!}{s^{n+1}},$$

where we have used (4.11).

4.2 The Existence of the Laplace Transform

In this section, we study carefully the assumptions under which a function f has a Laplace transform. In other words, under which assumptions on the function f the improper integral

$$\int_0^\infty f(t) e^{-st} \, dt \tag{4.12}$$

converges. For example if we take $f(t) = 1/t$, then the integral (4.12) does not converge at 0 since it grows too fast near zero. This is also the case with the function $f(t) = e^{t^2}$ which grows very fast at infinity and it cannot be controlled by e^{-st}.

It is well known from the theory of improper integrals that (4.12) converges if it converges absolutely. That is if the integral

$$\int_0^\infty |f(t) e^{-st}| \, dt \tag{4.13}$$

converges. Using the comparison test theorem in improper integrals, we then deduce that if there exists a function $g(t)$ such that

$$|f(t)e^{-st}| \leq g(t) \quad \text{and} \quad \int_0^{\infty} g(t)dt \tag{4.14}$$

converges, then (4.13) also converges and therefore (4.12) also does. The first requirement for this to happen is that the integral

$$\int_0^b f(t)e^{-st}dt$$

must exists for any finite number $b > 0$. To guarantee this, it is enough to assume that the function $f(t)e^{-st}$ is continuous or at least *piecewise continuous* as in Definition 4.2.1.

Definition 4.2.1 (Piecewise continuity)

A function $h(t)$ is said to be *piecewise continuous* on a finite interval $[a, b]$ if the interval $[a, b]$ can be partitioned into a finite number of subintervals

$$a = t_0 < t_1 < t_2 < \ldots < t_n = b$$

and the following two properties hold:
- the function $h(t)$ is continuous on each subinterval (t_i, t_{i+1}), $i = 0, \ldots, n-1$,
- the function $h(t)$ has a finite limit as t approaches each endpoint of each subinterval from its interior. i.e. the limits $\lim_{t \to t_i^+} h(t)$ and $\lim_{t \to t_{i+1}^-} h(t)$ exist as finite numbers.

A function $h(t)$ is said to be piecewise continuous on $[0, \infty)$ if $f(t)$ is piecewise continuous on $[0, b]$ for all $b > 0$.

Example 4.5
One of the simplest piecewise continuous function is the *unit step function* defined by

$$h(t) = \begin{cases} 1 & \text{for } t \geq 0, \\ 0 & \text{for } t < 0. \end{cases}$$

See ◘ Fig. 4.1.

If the function $f(t)$ is piecewise continuous for $t \geq 0$, then it follows that the integral

$$\int_0^b f(t)e^{-st}dt$$

exists for all $b < \infty$, but in order for $F(s)$ (the limit of this last integral as $b \to \infty$) to exist, we need some conditions on the behavior of the integrand $f(t)e^{-st}$ for large t. So, in order to ensure the convergence of the integral (4.12), then the function $f(t)$ should not grow too rapidly. Thus, we should assume that the function $f(t)$ is of *exponential order* as stated in the following definition.

□ **Fig. 4.1** The graph of the unite step function

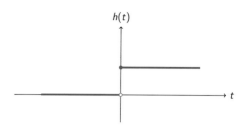

$h(t)$

t

Definition 4.2.2 (Exponential order)

A function $f(t)$ is said to be of *exponential order* α if there exist positive constants T and M such that

$$|f(t)| \leq Me^{\alpha t}, \tag{4.15}$$

for all $t \geq T$.

Example 4.6

It is clear that:

1. Any bounded function is of exponential order with $\alpha = 0$.
2. Any exponential function e^{at} is of exponential order with $\alpha = a$.
3. The function t^n is of exponential order with $\alpha > 0$, since for any $\alpha > 0$, and for any positive integer n,

$$t^n \leq Me^{\alpha t},$$

for some positive constant M.

4. The function e^{t^2} is not of exponential order since for any $\alpha > 0$ and for any positive constant $M \geq 1$, we have

$$e^{t^2} > Me^{\alpha t}, \quad \text{for} \quad t > T,$$

where $T = \frac{1}{2}(\alpha + (\alpha^2 + \ln M)^{1/2})$.

Theorem 4.2.1

If $f(t)$ is piecewise continuous on $[0, \infty)$ and of exponential order α, then its Laplace transform

$$\int_0^\infty f(t)e^{-st}\,dt$$

exists for $s > \alpha$.

Proof

Since $f(t)$ is of an exponential order, then (4.15) holds for some $M > 0$ and $T > 0$. Thus, we may split the above integral as

$$\int_0^\infty f(t)e^{-st}\,dt = \int_0^T f(t)e^{-st}\,dt + \int_T^\infty f(t)e^{-st}\,dt$$

$$= I_1 + I_2.$$

Since $f(t)$ is piecewise continuous, then the integral I_1 exists. To prove that I_2 converges, it is enough to show that (4.14) is satisfied. Indeed, since $f(t)$ is of exponential order, then for any $t \geq T$, we have

$$|f(t)e^{-st}| \leq Me^{-(s-\alpha)t}.$$

Thus, we have

$$\int_T^\infty |f(t)e^{-st}|\,dt \leq \int_T^\infty Me^{-(s-\alpha)t}\,dt = \frac{Me^{-(s-\alpha)T}}{s-\alpha},$$

since $s > \alpha$. Thus, I_2 converges absolutely, then it converges.

ℹ️ **Remark 4.2.2** Theorem 4.2.1 shows that any piecewise continuous function of exponential order has a Laplace transform. So, these assumptions are sufficient of the existence of the Laplace transform, but are not necessary. For example the function $f(t) = \frac{1}{\sqrt{t}}$ is not a piecewise continuous since it has an infinite discontinuity at $t = 0$, but its Laplace transform exists and we can prove that (see Exercise 4.4)

$$\mathcal{L}\{1/\sqrt{t}\}(s) = \sqrt{\pi/s}.$$

4.3   Properties of the Laplace Transform

In this section, we prove some properties of the Laplace transform that will be useful later.

4.3.1 Linearity

We start with one of the most important properties of the Laplace transform which is the linearity of the Laplace transform.

Theorem 4.3.1 (Linearity)
Let f, f_1 and f_2 be functions whose Laplace transforms exist for $s > \alpha$ and let λ be a constant, then for any $s > \alpha$, the following two properties hold:
1. $\mathcal{L}\{f_1 + f_2\} = \mathcal{L}\{f_1\} + \mathcal{L}\{f_2\}$,
2. $\mathcal{L}\{\lambda f\} = \lambda \mathcal{L}\{f\}$.

The proof of Theorem 4.3.1 is a direct application of the linearity of the integration. We omit the details.

Example 4.7

It is obvious that form the above property and Example 4.1 that for any constant a, then

$$\mathcal{L}\{a\}(s) = a\mathcal{L}\{1\}(s) = \frac{a}{s}, \qquad s > 0.$$

Also, using this last formula together with the first property in Theorem 4.3.1 and (4.10), then we can write the Laplace transform of any polynomial of order n.

Example 4.8

Show that

$$\mathcal{L}\{\cosh at\}(s) = \frac{s}{s^2 - a^2}, \qquad s > a \qquad\qquad (4.16)$$

and

$$\mathcal{L}\{\sinh at\}(s) = \frac{a}{s^2 - a^2}, \qquad s > a. \qquad\qquad (4.17)$$

Solution

We write $\cosh at$ and $\sinh at$ as

$$\cosh at = \frac{e^{at} + e^{-at}}{2}, \qquad \text{and} \qquad \sinh at = \frac{e^{at} - e^{-at}}{2}.$$

Using the linearity of the Laplace transform and Example 4.2, we get for $s > a$,

$$\mathcal{L}\{\cosh at\}(s) = \mathcal{L}\left\{\frac{e^{at} + e^{-at}}{2}\right\}(s)$$

$$= \frac{1}{2}\mathcal{L}\{e^{at}\}(s) + \frac{1}{2}\mathcal{L}\{e^{-at}\}(s)$$

$$= \frac{1}{2}\frac{1}{s-a} + \frac{1}{2}\frac{1}{s+a}$$

$$= \frac{s}{s^2 - a^2},$$

which proves (4.16). Similarly, we have for $s > a$,

$$\mathcal{L}\{\sinh at\}(s) = \mathcal{L}\left\{\frac{e^{at} - e^{-at}}{2}\right\}(s)$$

$$= \frac{1}{2}\frac{1}{s-a} - \frac{1}{2}\frac{1}{s+a}$$

$$= \frac{a}{s^2 - a^2},$$

which gives (4.17).

4.3.2 Translation

We have proved in Lemma 3.3.4 the exponential shift formula. We show here that a similar formula holds for the Laplace transform.

Theorem 4.3.2 (Translation formula)
Assume that the Laplace transform of a function f: $\mathcal{L}\{f\}(s) = F(s)$ exists for $s > \alpha$. Then, we have

$$\mathcal{L}\{e^{at} f\}(s) = F(s - a), \quad \text{for} \quad s > \alpha + a. \tag{4.18}$$

Proof
The proof of (4.18) is a direct application of Definition 4.1.1. Indeed, we have

$$\mathcal{L}\{e^{at} f\}(s) = \int_0^\infty e^{at} f(t) e^{-st} dt$$

$$= \int_0^\infty f(t) e^{-(s-a)t} dt$$

$$= F(s - a),$$

for $s - a > 0$.

Example 4.9
Use (4.18) to find the Laplace transform of the functions

$$f(t) = e^{at} \sin bt \quad \text{and} \quad g(t) = e^{at} \cos bt.$$

Solution
Using (4.18) and (4.8), we obtain

$$\mathcal{L}\{e^{at} \sin bt\}(s) = \frac{b}{(s-a)^2 + b^2}, \quad s > a.$$

Similarly, exploiting (4.18) and (4.9), we find

$$\mathcal{L}\{e^{at} \cos bt\}(s) = \frac{s-a}{(s-a)^2 + b^2}, \quad s > a.$$

Example 4.10
Find the Laplace transform of the function

$$f(t) = e^{at} t^n, \quad n = 0, 1, 2, \dots$$

Solution
It is clear from Example 4.4 that

$$F(s) = \mathcal{L}\{t^n\}(s) = \frac{n!}{s^{n+1}}, \quad n = 0, 1, 2, \dots \quad s > 0.$$

Now, applying formula (4.18), we find it clear from Example 4.4 that

$$\mathcal{L}\{e^{at} t^n\}(s) = F(s - a) = \frac{n!}{(s-a)^{n+1}}, \quad n = 0, 1, 2, \dots \quad s > a.$$

4.3.3 Derivative

Since, we will be later using the Laplace transform to solve differential equations, then, we show next, an important property of the Laplace transform which gives a relation between the Laplace transform of a function and the Laplace transform of its derivatives.

Theorem 4.3.3 (The Laplace transform of the derivative)
Let $f(t)$ be a differentiable function on $[0, \infty)$ with both $f(t)$ and $f'(t)$ are of exponential order α. Then, for $s > \alpha$,

$$\mathcal{L}\{f'\}(s) = s\mathcal{L}\{f\}(s) - f(0). \tag{4.19}$$

Proof
Using the definition of the Laplace transform, we get

$$\mathcal{L}\{f'\}(s) = \int_0^\infty f'(t)e^{-st}dt. \tag{4.20}$$

Now, using the integration by parts, we can write the right-hand side of (4.20) as

$$\int_0^\infty f'(t)e^{-st}dt = \left[e^{-st}f(t)\right]_0^\infty + s\int_0^\infty f(t)e^{-st}dt. \tag{4.21}$$

Now, we have

$$\left[e^{-st}f(t)\right]_0^\infty = \lim_{N \to \infty}\left[e^{-st}f(t)\right]_0^N$$
$$= \lim_{N \to \infty}\{e^{-st}f(t)\} - f(0).$$

Since $f(t)$ is of an exponential order α, then for any $s \geq \alpha$, we have $\lim_{N \to \infty}\{e^{-st}f(t)\} = 0$. Consequently, formula (4.21) takes the form:

$$\mathcal{L}\{f'\}(s) = -f(0) + s\int_0^\infty f(t)e^{-st}dt,$$

which is exactly (4.19).

Now, using induction, then we can extend the above formula (4.19) to higher derivatives of $f(t)$. For instance if $f(t)$ is differentiable twice on $[0, \infty)$ and $f(t)$, $f'(t)$ and $f''(t)$ are of exponential order α on $[0, \infty)$, then we have for all $s > \alpha$,

$$\mathcal{L}\{f''\}(s) = s\mathcal{L}\{f'\}(s) - f'(0)$$
$$= s[s\mathcal{L}\{f\}(s) - f(0)] - f'(0)$$
$$= s^2\mathcal{L}\{f\}(s) - sf(0) - f'(0). \tag{4.22}$$

By the same method, we may compute the Laplace transform of higher-order derivatives of a function $f(t)$ as in the following theorem.

Theorem 4.3.4 (Laplace transform of higher-order derivatives)
Let $f(t)$ be continuous on $[0, \infty)$ and assume that $f'(t), \ldots, f^{(n)}(t)$ exist and let $f^{(n)}(t)$ be piecewise continuous on $[0, \infty)$ and assume that all these functions are of exponential order α. Then, for all $s > \alpha$, we have

$$\mathcal{L}\{f^{(n)}\}(s) = s^n \mathcal{L}\{f\}(s) - s^{n-1} f(0) - s^{n-2} f'(0) - \ldots - f^{(n-1)}(0). \tag{4.23}$$

Example 4.11
Use formula (4.8) and (4.19) to prove (4.9).

Solution
Taking $f(t) = \sin bt$, we get, in one hand:

$$\mathcal{L}\{f'\}(s) = \mathcal{L}\{b \cos bt\}(s)$$
$$= b\mathcal{L}\{\cos bt\}(s). \tag{4.24}$$

On the other hand, we have, according to formula (4.19), that

$$\mathcal{L}\{f'\}(s) = s\mathcal{L}\{\sin t\}(s) - \sin(0)$$
$$= s\mathcal{L}\{\sin t\}(s)$$
$$= \frac{bs}{s^2 + b^2}, \quad s > 0. \tag{4.25}$$

Thus, combining (4.24) and (4.25), we find (4.9).

Let us consider the general formula of the Laplace transform

$$F(s) = \int_0^\infty e^{-st} f(t)dt.$$

Now, we apply the Leibniz formula for integration, then we get

$$\frac{d}{ds} F(s) = \int_0^\infty \frac{\partial}{\partial s}[e^{-st} f(t)]dt$$
$$= -\int_0^\infty te^{-st} f(t)dt,$$

which can be written as

$$\mathcal{L}\{tf(t)\}(s) = -\frac{d}{ds} F(s). \tag{4.26}$$

We can simply use induction and extend formula (4.26) to the nth derivative as in the following theorem.

Theorem 4.3.5

Let $f(t)$ be a piecewise continuous function on $[0, \infty)$ and of exponential order α and let $F(s) = \mathcal{L}\{f\}(s)$. Then for any $s > \alpha$, we have

$$\mathcal{L}\{t^n f\}(s) = (-1)^n \frac{d^n}{ds^n} F(s).$$ (4.27)

Example 4.12

Find the Laplace transform of $f(t) = t \cos bt$.

Solution

Using (4.9) and formula (4.27), we obtain

$$\mathcal{L}\{t \cos bt\}(s) = -\frac{d}{ds} F(s),$$

where

$$F(s) = \mathcal{L}\{\cos bt\}(s) = \frac{s}{s^2 + b^2}.$$

Thus, we have from above that

$$\mathcal{L}\{t \cos bt\}(s) = \frac{s^2 - b^2}{(s^2 + b^2)^2}, \quad s > 0.$$

4.3.4 Integration

Now, we give a formula for the integration of Laplace transform. Thus, we have the following theorem.

Theorem 4.3.6

Let $f(t)$ be a piecewise continuous function on $[0, \infty)$ and of exponential order α and let $F(s) = \mathcal{L}\{f\}(s)$. Then,

$$\int_s^\infty F(s)ds = \mathcal{L}\left\{\frac{f(t)}{t}\right\}(s).$$ (4.28)

Proof

We put $G(s) = \mathcal{L}\{f(t)/t\}(s)$. Then, a simple application of (4.27) for $n = 1$, leads to

$$\frac{dG(s)}{ds} = \mathcal{L}\left\{(-t)\frac{f(t)}{t}\right\}(s)$$
$$= -\mathcal{L}\{f(t)\}(s)$$
$$= -F(s).$$

Therefore,

$$G(s) = - \int_a^s F(s)ds \qquad (4.29)$$

for some a. Since $G(s)$ is the Laplace transform of some function, then $\lim_{s \to \infty} G(s) = 0$. So, in order to get this last property, then we need to put $a = \infty$ in (4.29) and then, we get

$$G(s) = \int_s^\infty F(s)ds,$$

which is exactly (4.28).

Remark 4.3.7 Formula (4.28) can be rewritten as

$$\int_0^\infty e^{-st} \frac{f(t)}{t} dt = \int_s^\infty F(s)ds.$$

Then, by letting $s \to 0$, we get

$$\int_0^\infty \frac{f(t)}{t} dt = \int_0^\infty F(s)ds, \qquad (4.30)$$

which is valid whenever the integral on the left exists.

Formula (4.30) can sometimes be used to evaluate integrals that are difficult to handle by other methods.

Example 4.13
Show that

$$\int_0^\infty \frac{\sin t}{t} dt = \frac{\pi}{2}. \qquad (4.31)$$

Solution
Using formula (4.30), we get

$$\int_0^\infty \frac{\sin t}{t} dt = \int_0^\infty F(s)ds,$$

where $F(s) = \frac{1}{s^2+1}$ is the Laplace transform of the function $f(t) = \sin t$. Consequently, we obtain

$$\int_0^\infty \frac{\sin t}{t} dt = \int_0^\infty \frac{1}{s^2+1} ds = \tan^{-1} s|_{s=0}^\infty = \frac{\pi}{2}.$$

4.3.5 Convolution

In this subsection, we give another important property of the Laplace transform. Let $F(s)$ be the Laplace transform of a function $f(t)$ and $G(s)$ be the Laplace transform of $g(t)$. The question that we should ask is the following: *can we find a formula for the Laplace transform of $f(t)g(t)$ depending on $F(s)$ and $G(s)$?* The answer to the above question is no. So, let us reverse the question as follows: *can we write the product $F(s)G(s)$ as the Laplace transform of a function depending on f and g?* Now, the answer is yes. In fact $F(s)G(s)$ is the Laplace transform of the *convolution* of $f(t)$ and $g(t)$.

Definition 4.3.1 (Convolution)

Let $f(t)$ and $g(t)$ be piecewise continuous functions on $[0, \infty)$. Then, the *convolution* of $f(t)$ and $g(t)$, is denoted by $f * g$ and defined as

$$(f * g)(t) = \int_0^t f(u)g(t - u)du. \tag{4.32}$$

Example 4.14

Find the convolution of $f(t) = \cos t$ and $g(t) = \sin t$.

Solution

We have

$$(f * g)(t) = \int_0^t \cos u \sin(t - u)du.$$

Using the formula

$$\cos \alpha \sin \beta = \frac{1}{2} \sin(\alpha + \beta) + \frac{1}{2} \sin(\alpha - \beta),$$

we obtain

$$\int_0^t \cos u \sin(t - u)du = \frac{1}{2} \int_0^t \left(\sin t + \sin(t - 2u) \right) du$$

$$= \frac{1}{2} \left[u \sin t + \frac{1}{2} \cos(t - 2u) \right]_{u=0}^{u=t}$$

$$= \frac{1}{2} t \sin t.$$

Example 4.15

It is clear that if we take $g(t) = 1$ for all $t > 0$, then we get from (4.32) that

$$(f * 1)(t) = \int_0^t f(u)du.$$

Theorem 4.3.8

Let $f(t)$ and $g(t)$ be piecewise continuous functions on $[0, \infty)$. Then for all $t > 0$,

$$(f * g)(t) = (g * f)(t).$$

Proof

We have

$$(f * g)(t) = \int_0^t f(u)g(t-u)du. \tag{4.33}$$

Now, we put the change of variable $v = t - u$, then the integral in (4.33) can be rewritten as

$$(f * g)(t) = \int_t^0 f(t-v)g(v)(-dv)$$

$$= \int_0^t f(t-v)g(v)dv$$

$$= (g * f)(t).$$

This completes the proof of Theorem 4.3.8.

Next, we introduce an important property of the Laplace transform of the convolution of two functions. We have seen in ▶ Sect. 4.1 that the Laplace transform is the continuous analogs of power series.

It is known from power series that if $A(x) = \sum_{n=0}^{\infty} a(n)x^n$ and $B(x) = \sum_{n=0}^{\infty} b(n)x^n$ are two series that converge absolutely for $|x| < r$, then

$$\left(\sum_{n=0}^{\infty} a(n)x^n \right) \cdot \left(\sum_{n=0}^{\infty} b(n)x^n \right) = \sum_{n=0}^{\infty} c(n)x^n, \tag{4.34}$$

with

$$c(n) = \sum_{k=0}^{n} a(k)b(n-k)$$

converges absolutely to $A(x)B(x)$ for $|x| < r$. Thus,

$$A(x)B(x) = \sum_{n=0}^{\infty} c(n)x^n. \tag{4.35}$$

The left-hand side in (4.35) is the discrete analogs of the convolution (4.32). In the next theorem we prove a formula similar to (4.35).

Theorem 4.3.9

Let $f(t)$ and $g(t)$ be piecewise continuous functions on $[0, \infty)$ and of exponential order α and let $F(s) = \mathcal{L}\{f\}(s)$ and $G(s) = \mathcal{L}\{g\}(s)$. Then, we have

$$\mathcal{L}\{(f * g)\}(s) = F(s)G(s). \tag{4.36}$$

Proof

To prove (4.36), we need some tools for the change of variables in double integrals. Let us start first computing the right-hand side of (4.36)

$$F(s)G(s) = \left(\int_0^\infty e^{-su} f(u)du \right) \cdot \left(\int_0^\infty e^{-sv} g(v)dv \right)$$

$$= \int_0^\infty \int_0^\infty e^{-s(u+v)} f(u)g(v)dudv. \tag{4.37}$$

Now, we put $u + v = t$ which gives $v = t - u$. Next, we need to change form the variables u and v in (4.37) to the new variables

$$\begin{cases} u = u, \\ v = t - u. \end{cases}$$

Thus, we write $dudv$ in terms of $dudt$ as

$$dudv = \det \begin{bmatrix} \frac{\partial u}{\partial u} & \frac{\partial u}{\partial t} \\ \frac{\partial v}{\partial u} & \frac{\partial v}{\partial t} \end{bmatrix} dudt = \det \begin{bmatrix} 1 & 0 \\ -1 & 1 \end{bmatrix} dudt$$

$$= dudt.$$

Now, to find the new limits of the integral, we first let t to be fixed and vary u. We have t fixed, means that $u + v$ is fixed and u varies from $u = 0$ to $v = 0$ which means $u = t$. For u fixed, then t changes from 0 to ∞. Consequently, formula (4.37) becomes

$$F(s)G(s) = \int_0^\infty \int_0^t e^{-st} f(u)g(t-u)dudt$$

$$= \int_0^\infty e^{-st} \left(\int_0^t f(u)g(t-u)du \right) dt$$

$$= \mathcal{L}\{(f * g)\}(s).$$

This finishes the proof of Theorem 4.3.9.

Example 4.16

Find the solution of the integral equation

$$y(t) = 1 - \int_0^t y(u)(t-u)du. \tag{4.38}$$

Solution

Equation (4.38) can be written as

$$y(t) = 1 - (y * t)(t). \tag{4.39}$$

Applying the Laplace transform to (4.39) and making use of (4.36), we obtain

$$\begin{aligned}
Y(s) &= \mathcal{L}\{1\}(s) - \mathcal{L}\{(y * t)\}(s) \\
&= \frac{1}{s} - Y(s)\mathcal{L}\{t\}(s) \\
&= \frac{1}{s} - Y(s)\frac{1}{s^2},
\end{aligned}$$

where $Y(s) = \mathcal{L}\{y(t)\}(s)$. Arranging the above equation, we find that

$$Y(s) = \frac{s}{s^2 + 1}, \quad s > 0. \tag{4.40}$$

Thus, the solution of (4.39) is the function $y(t)$ such that its Laplace transform satisfies (4.40). Recalling (4.9), we deduce that the solution of (4.39) is

$$y(t) = \cos t.$$

4.3.6 Laplace Transform of a Periodic Function

In this subsection, we give the Laplace transform of a periodic function.

> **Definition 4.3.2 (Periodic function)**
>
> A function $f(t)$ is said to be *periodic* of period $T \neq 0$ if
>
> $$f(t + T) = f(t),$$
>
> for all t in the domain of f.

In the next theorem, we introduce the formula of the Laplace transform of a periodic function.

Theorem 4.3.10

Let $f(t)$ be a piecewise continuous function on $[0, T]$ and periodic of period T. Then its Laplace transform $F(s)$ is given by

$$F(s) = \frac{F_T(s)}{1 - e^{-st}}, \tag{4.41}$$

where

$$F_T(s) = \int_0^T e^{-sT} f(t)dt.$$

Proof

Since $f(t)$ is periodic of a period T, then we may write its Laplace transform as

$$F(s) = \lim_{n\to\infty} \int_0^T e^{-sT} f(t)dt + \int_T^{2T} e^{-sT} f(t)dt + \ldots + \int_{nT}^{(n+1)T} e^{-sT} f(t)dt + \ldots$$

Now, we make the change of variable $u = t - T$ in the second integral, …, the change of variable $u = t - nT$ in the $(n+1)$st integral, we obtain

$$F(s) = \lim_{n\to\infty} (1 + e^{-sT} + \ldots + e^{-snT}) \int_0^T e^{-sT} f(t)dt. \tag{4.42}$$

The term

$$1 + e^{-sT} + \ldots + e^{-snT}$$

is the partial sum of order n of a geometric series which is converging since $|e^{-sT}| < 1$, for $s > 0$ and $T \neq 0$. Consequently

$$\lim_{n\to\infty} (1 + e^{-sT} + \ldots + e^{-snT}) = \frac{1}{1 - e^{-sT}}.$$

Thus, plugging this last formula into (4.42), we get (4.41).

□ Table 4.1 Brief table for Laplace transform

$f(t)$	$F(s) = \mathcal{L}\{f(t)\}(s)$	
1	$\dfrac{1}{s}$	$s > 0$
e^{at}	$\dfrac{1}{s-a}$	$s > a$
$t^n, \quad n = 0, 1, 2, \ldots$	$\dfrac{n!}{s^{n+1}}$	$s > 0$
$\sin bt$	$\dfrac{b}{s^2 + b^2},$	$s > 0$
$\cos bt$	$\dfrac{s}{s^2 + b^2},$	$s > 0$
$t^n e^{at}, \quad n = 0, 1, 2, \ldots$	$\dfrac{n!}{(s-a)^{n+1}}$	$s > a$
$e^{at} \sin bt$	$\dfrac{b}{(s-a)^2 + b^2}$	$s > a$
$e^{at} \cos bt$	$\dfrac{s-a}{(s-a)^2 + b^2}$	$s > a$
$e^{at} f(t)$	$F(s-a)$	
$\displaystyle\int_0^s f(t)dt$	$\dfrac{1}{s} F(s)$	

4.4   Inverse Laplace Transform

As we have said before, Laplace transform is an alternative and a simple way of solving differential equations. Let us for instance consider the initial value problem

$$\begin{cases} ay'' + by' + cy = f(t), \\ y(0) = y_0, \quad y'(0) = y_1. \end{cases} \tag{4.43}$$

To solve (4.43), let us assume that $Y(s)$ is the Laplace transform of the solution $y(t)$ and $F(s)$ is the Laplace transform of $f(t)$. Thus, taking the Laplace transform of both sides in the first equation of (4.43) and using the linearity of the Laplace transform and the formula of the Laplace transform of the derivatives, we get

$$a \left(s^2 Y(s) - s y_0 - y_1 \right) + b \left(s Y(s) - y_0 \right) + c Y(s) = F(s).$$

This gives

$$Y(s) = \frac{F(s) + (as + b) y_0 + a y_1}{as^2 + bs + c}. \tag{4.44}$$

Now, in order to find the solution $y(t)$ of (4.43), we need to find the function $y(t)$ that has the right-hand side in (4.44) as its Laplace transform. The procedure of finding $y(t)$ from $Y(s)$ is called the *inverse Laplace transform* of $Y(s)$.

Definition 4.4.1 (Inverse Laplace transform)

Given a function $F(s)$ such that $\lim_{s \to \infty} F(s) = 0^1$. If there is a function $f(t)$ that is continuous on $[0, \infty)$ and satisfies

$$\mathcal{L}\{f\}(s) = F(s),$$

then we say that $f(t)$ is the *inverse Laplace transform* of $F(s)$ and employ the notation

$$\mathcal{L}^{-1}\{F(s)\} = f(t).$$

Example 4.17

Find the inverse Laplace transform of the following functions:

1. $F(s) = \dfrac{s}{s^2 + 4}$,

2. $F(s) = \dfrac{1}{s^2}$,

3. $F(s) = \dfrac{6}{(s + 2)^2 + 9}$.

---

[1] It can be shown that if $F(s)$ is a function of s with the property that its limit as $s \to \infty$ does not exist or is not equal to zero, then it cannot be the Laplace transform of any function $f(t)$. For example, polynomials in s, $\sin s$, $\cos s$, e^s, and $\ln s$ cannot be Laplace transforms of other functions.

Solution

1. Using ■ Table 4.1, we get:

$$\mathcal{L}^{-1}\left\{\frac{s}{s^2+4}\right\} = \cos 2t.$$

2. Similarly, we have by using ■ Table 4.1

$$\mathcal{L}^{-1}\left\{\frac{1}{s^2}\right\} = t.$$

3. Also,

$$\mathcal{L}^{-1}\left\{\frac{6}{(s+2)^2+9}\right\} = 2e^{-2t}\sin 3t.$$

One of the most useful properties of the inverse Laplace transform is the linearity of this transform.

Theorem 4.4.1

Assume that $\mathcal{L}^{-1}\{F\}$, $\mathcal{L}^{-1}\{F_1\}$ and $\mathcal{L}^{-1}\{F_2\}$ exist and are continuous on $[0, \infty)$ and let λ be a constant. Then, we have

1. $\mathcal{L}^{-1}\{F_1 + F_2\} = \mathcal{L}^{-1}\{F_1\} + \mathcal{L}^{-1}\{F_2\}$;
2. $\mathcal{L}^{-1}\{\lambda F\} = \lambda \mathcal{L}^{-1}\{F\}$.

The proof of Theorem 4.4.1 is a direct consequence of Theorem 4.3.1, we omit it.

Example 4.18

Find the inverse Laplace transform of the function

$$F(s) = \frac{2}{s-1} + \frac{3}{s+1} + \frac{4s}{s^2+4}. \tag{4.45}$$

Solution

Using the linearity of the inverse Laplace transform, we obtain

$$\mathcal{L}^{-1}\{F(s)\} = \mathcal{L}^{-1}\left\{\frac{2}{s-1}\right\} + \mathcal{L}^{-1}\left\{\frac{3}{s+1}\right\} + \mathcal{L}^{-1}\left\{\frac{4s}{s^2+4}\right\}$$

$$= 2\mathcal{L}^{-1}\left\{\frac{1}{s-1}\right\} + 3\mathcal{L}^{-1}\left\{\frac{1}{s+1}\right\} + 4\mathcal{L}^{-1}\left\{\frac{s}{s^2+4}\right\}$$

$$= 2e^t + 3e^{-t} + 4\cos 2t,$$

where we have used ■ Table 4.1.

Example 4.19

Find the inverse Laplace transform of the function

$$F(s) = \frac{1}{(s+1)^3} + \frac{3}{s^2+2s+2}. \tag{4.46}$$

Solution

The linearity of the inverse Laplace transform gives

$$\mathcal{L}^{-1}\{F(s)\} = \mathcal{L}^{-1}\left\{\frac{1}{(s+1)^3}\right\} + \mathcal{L}^{-1}\left\{\frac{3}{s^2+2s+2}\right\}. \tag{4.47}$$

Since

$$\mathcal{L}\{t^n e^{at}\} = \frac{n!}{(s-a)^{n+1}},$$

then, we get, by taking $n = 2$,

$$\mathcal{L}^{-1}\left\{\frac{1}{(s+1)^3}\right\} = \frac{1}{2}t^2 e^{-t}.$$

On the other hand, to find the second term on the right-hand side of (4.47), we need to complete the square of the denominator and write

$$\mathcal{L}^{-1}\left\{\frac{3}{s^2+2s+2}\right\} = 3\mathcal{L}^{-1}\left\{\frac{1}{(s+1)^2+1}\right\}.$$

Since

$$\mathcal{L}\{e^{at}\sin bt\} = \frac{b}{(s-a)^2+b^2},$$

then, we get with $a = -1$ and $b = 1$,

$$\mathcal{L}^{-1}\left\{\frac{1}{(s+1)^2+1}\right\} = e^{-t}\sin t.$$

Collecting the above formulas, we arrive at

$$\mathcal{L}^{-1}\left\{\frac{1}{(s+1)^3} + \frac{3}{s^2+2s+2}\right\} = \frac{1}{2}t^2 e^{-t} + 3e^{-t}\sin t.$$

Example 4.20

Find the inverse Laplace transform of

$$F(s) = \frac{5s-3}{s^2-2s-3}. \tag{4.48}$$

Solution

If we rely only on the Laplace transform table, it is not obvious how we can get the inverse Laplace transform of $F(s)$ in (4.48). Thus, before using ◻ Table 4.1, let us first rewrite $F(s)$ as

$$F(s) = \frac{2}{s+1} + \frac{3}{s-3}. \tag{4.49}$$

Therefore, using the linearity of the inverse Laplace transform and ◻ Table 4.1, we obtain

$$\mathcal{L}^{-1}\{F(s)\} = 2\mathcal{L}^{-1}\left\{\frac{1}{s+1}\right\} + 3\mathcal{L}^{-1}\left\{\frac{1}{s-3}\right\}$$
$$= 2e^{-t} + 3e^{3t}.$$

The method that we used to obtain (4.49) is known as the *method of partial factions*. Let us describe this method in more general context.

General Description of the Method of Partial Fractions Assume that $F(s) = h(s)/g(s)$ such that $h(s)$ and $g(s)$ are two polynomials and the degree of $h(s)$ is strictly less than the degree of $g(s)$. Then, we may write a new form of $F(s)$ as the sum of partial fractions following the steps:

1. Let $s - r$ be a linear factor of $g(s)$. Suppose that $(s - r)^m$ is the highest power of $s - r$ that divides $g(s)$. Then to this factor, assign the sum of the m partial fractions:

$$\frac{A_1}{s - r} + \frac{A_2}{(s - r)^2} + \ldots + \frac{A_m}{(s - r)^m}.$$

Do this for each distinct linear factor of $g(s)$.

2. Let $s^2 + ps + q$ be a quadratic factor of $g(s)$. Suppose that $(s^2 + ps + q)^n$ is the highest power of this factor that divides $g(s)$. Then, to this factor assign the sum of the n partial fractions:

$$\frac{B_1 s + C_1}{s^2 + ps + q} + \frac{B_2 s + C_2}{(s^2 + ps + q)^2} + \ldots + \frac{B_n s + C_n}{(s^2 + ps + q)^n}.$$

Do this for each distinct quadratic factor of $g(s)$ that cannot be factored into linear factors with real coefficients.

3. Set $F(s)$ equal to the sum of all these partial fractions and then find the constants.

Example 4.21

Find the inverse Laplace transform of the function

$$F(s) = \frac{s^2 + 4s + 1}{(s - 1)(s + 1)(s + 3)}.$$

Solution

Using the above method, we may write $F(s)$ as follows:

$$F(s) = \frac{A}{s - 1} + \frac{B}{s + 1} + \frac{C}{s + 3}. \tag{4.50}$$

The easiest method to find the constants A, B and C in (4.50) is a method know as the *Heaviside method* which says in order to find A in (4.50), we need to multiply both sides by $(s - 1)$ and take $s = 1$, this gives $A = 3/4$. Similarly, multiplying both sides in (4.50) by $(s + 1)$ and take $s = -1$, we get $B = 1/2$. To find C, we multiply now by $(s + 3)$ and take $s = -3$, we obtain $C = -1/4$. Consequently, we have

$$F(s) = \frac{1}{4(s - 1)} + \frac{1}{2(s + 1)} - \frac{1}{4(s + 3)}.$$

Therefore,

$$\mathcal{L}^{-1}\{F(s)\} = \frac{1}{4}\mathcal{L}^{-1}\left\{\frac{1}{s - 1}\right\} + \frac{1}{2}\mathcal{L}^{-1}\left\{\frac{1}{s + 1}\right\} - \frac{1}{4}\mathcal{L}^{-1}\left\{\frac{1}{s + 3}\right\}$$

$$= \frac{1}{4}e^t + \frac{1}{2}e^t - \frac{1}{4}e^{-3t}.$$

Example 4.22

Find the inverse Laplace transform of the function

$$F(s) = \frac{6s + 7}{(s + 2)^2}.$$

Solution

First, we express $F(s)$ as a sum of partial fractions as:

$$F(s) = \frac{A}{s + 2} + \frac{B}{(s + 2)^2}.$$

To find the constants A and B, we first multiply both sides in the above equation by $(s + 2)^2$, to obtain

$$6s + 7 = A(s + 2) + B. \qquad (4.51)$$

By taking $s = -2$, we find $B = -5$. Next, we differentiate both sides in (4.51), with respect to s, to get $A = 6$. Consequently, $F(s)$ can be written as

$$F(s) = \frac{6}{s + 2} - \frac{5}{(s + 2)^2}.$$

Hence,

$$\mathcal{L}^{-1}\{F(s)\} = 6\mathcal{L}^{-1}\left\{\frac{1}{s + 2}\right\} - 5\mathcal{L}^{-1}\left\{\frac{1}{(s + 2)^2}\right\}$$
$$= 6e^{-2t} - 5te^{-2t}.$$

Example 4.23

Find the inverse Laplace transform of the function

$$F(s) = \frac{-2s + 4}{(s^2 + 1)(s - 1)^2}.$$

Solution

The denominator in $F(s)$ has an irreducible quadratic factor as well as a repeated linear factor, so, we can write $F(s)$ as

$$F(s) = \frac{As + B}{s^2 + 1} + \frac{C}{s - 1} + \frac{D}{(s - 1)^2}. \qquad (4.52)$$

We may find the constants in the above formula as follows: multiplying formula (4.52) by $(s^2 + 1)(s^2 - 1)$, we get

$$-2s + 4 = (As + B)(s - 1)^2 + C(s - 1)(s^2 + 1) + D(s^2 + 1).$$

This gives

$$-2s + 4 = (A + C)s^3 + (-2A + B - C + D)s^2 + (A - 2B + C)s + (B - C + D).$$

Thus, by equating the coefficients of like powers of s, we get the following system of algebraic equations

$$\begin{cases} A + C = 0, \\ -2A + B - C + D = 0, \\ A - 2B + C = -2, \\ B - C + D = 4. \end{cases}$$

Solving the above system, we find $A = 2$, $B = 1$, $C = -2$ and $D = 1$.

Plugging these into (4.52), we obtain

$$F(s) = \frac{2s+1}{s^2+1} - \frac{2}{s-1} + \frac{1}{(s-1)^2}$$

$$= \frac{2s}{s^2+1} + \frac{1}{s^2+1} - \frac{2}{s-1} + \frac{1}{(s-1)^2}.$$

Therefore, we get

$$\mathcal{L}^{-1}\{F(s)\} = 2\mathcal{L}^{-1}\left\{\frac{s}{s^2+1}\right\} + \mathcal{L}^{-1}\left\{\frac{1}{s^2+1}\right\} - 2\mathcal{L}^{-1}\left\{\frac{1}{s-1}\right\} + \mathcal{L}^{-1}\left\{\frac{1}{(s-1)^2}\right\}$$

$$= 2\cos t + \sin t - 2e^t + te^t.$$

Example 4.24

Write the inverse Laplace transform of the function

$$F(s) = \frac{1}{s(s^2+1)^2}.$$

Solution

The function $F(s)$ has a repeated irreducible quadratic factor, thus its partial fraction decomposition is

$$F(s) = \frac{A}{s} + \frac{Bs+C}{s^2+1} + \frac{Ds+E}{(s^2+1)^2}.$$

Multiplying the above formula by $s(s^2+1)^2$, we obtain

$$1 = A(s^2+1)^2 + (Bs+C)s(s^2+1) + (Ds+E)s.$$

This gives after a simple computation

$$1 = (A+B)s^4 + Cs^3 + (2A+B+D)s^2 + (C+E)s + A,$$

which leads to the following system of equations

$$\begin{cases} A+B = 0, \\ C = 0, \\ 2A+B+D = 0, \\ C+E = 0, \\ A = 1. \end{cases}$$

The solution of the above system is $A = 1$, $B = -1$, $C = 0$, $D = -1$ and $E = 0$. Consequently, plugging these into the above formula of $F(s)$, we get

$$F(s) = \frac{1}{s} - \frac{s}{s^2+1} - \frac{s}{(s^2+1)^2}.$$

Now, we have

$$\mathcal{L}^{-1}\{F(s)\} = \mathcal{L}^{-1}\left\{\frac{1}{s}\right\} - \mathcal{L}^{-1}\left\{\frac{s}{s^2+1}\right\} - \mathcal{L}^{-1}\left\{\frac{s}{(s^2+1)^2}\right\}.$$

Using ▣ Table 4.1, we have

$$\mathcal{L}^{-1}\left\{\frac{1}{s}\right\} = 1.$$

and

$$\mathcal{L}^{-1}\left\{\frac{s}{s^2+1}\right\} = \cos t.$$

On the other hand, using (4.27), we get

$$\mathcal{L}\{t \sin bt\} = -\frac{d}{ds}\left(\frac{b}{s^2+b^2}\right) = \frac{2sb}{(s^2+b^2)^2}.$$

Applying the above formula for $b = 1$, we find

$$\mathcal{L}^{-1}\left\{\frac{s}{(s^2+1)^2}\right\} = \frac{1}{2}t \sin t.$$

Finally, we have

$$\mathcal{L}^{-1}\{F(s)\} = 1 - \cos t - \frac{1}{2}t \sin t.$$

4.5   Applications to Differential Equations

As we have said earlier, Laplace transform is one of the powerful tools for solving differential equations. Due to the appearance of the initial values in the Laplace transform of the derivatives, then the Laplace transform method can be used to solve differential equations with initial values (initial value problems). The advantage here is that this method leads directly to the solution of the initial value problem rather than finding the general solution and then using the initial values for obtaining the constants of the integration. In addition, and as we will see later, this method can easily handle equations involving forcing functions having jump discontinuities. Furthermore, the method can be also used for solving differential equations with variable coefficients, integral equations and systems of differential equations.

In order to solve an initial value problem using the Laplace transform method, we need to do the following three main steps:

1. Take the Laplace transform of both sides in the equation.
2. Use the properties of the Laplace transform and the initial values to obtain an algebraic equation for the Laplace transform of the solution and then solve this equation for the transform.
3. To get the solution of the initial value problem, we need just to get the inverse Laplace transform.

Example 4.25
Find the solution of the initial value problem

$$\begin{cases} y''(t) + 2y'(t) + y(t) = e^{-t}, \\ y(0) = 0, \qquad y'(0) = 0. \end{cases} \tag{4.53}$$

Solution
Taking the Laplace transform of the first equation in (4.53), we get

$$s^2 Y(s) - y'(0) - sy(0) + 2\Big(sY(s) - y(0)\Big) + Y(s) = \mathcal{L}\{e^{-t}\},$$

where $Y(s) = \mathcal{L}\{y\}(s)$. Making use of the initial values and ◨ Table 4.1, we obtain

$$(s^2 + 2s + 1)Y(s) = \frac{1}{1+s}.$$

Solving the above equation, we find

$$Y(s) = \frac{1}{(s+1)^3}.$$

We have seen in Example 4.19 that

$$\mathcal{L}^{-1}\left\{\frac{1}{(s+1)^3}\right\} = \frac{1}{2}t^2 e^{-t}.$$

Consequently, the solution of (4.53) is

$$y(t) = \frac{1}{2}t^2 e^{-t}.$$

4.6 Exercises

Exercise 4.1
Find the solution to the initial value problem

$$\begin{cases} x'(t) = 2x(t) + 5y(t), \\ y'(t) = x(t) - 2y(t), \\ x(0) = 1, \quad y(0) = 0. \end{cases} \tag{4.54}$$

Solution
Applying the Laplace transform to the first two equations in (4.54), we find for $X(s) = \mathcal{L}\{x\}(s)$ and $Y(s) = \mathcal{L}\{y\}(s)$,

$$\begin{cases} sX(s) - x(0) = 2X(s) + 5Y(s), \\ sY(s) - y(0) = X(s) - 2Y(s). \end{cases}$$

This gives, by using the initial values

$$\begin{cases} (s-2)X(s) - 5Y(s) = 1, \\ -X(s) + (s+2)Y(s) = 0. \end{cases} \tag{4.55}$$

Solving the above algebraic system for the unknowns $X(s)$ and $Y(s)$, we get

$$X(s) = \frac{s+2}{s^2-9}, \quad \text{and} \quad Y(s) = \frac{1}{s^2-9}.$$

Using the method of partial fractions, we can write $X(s)$ and $Y(s)$ as

$$X(s) = \frac{5}{6(s-3)} + \frac{1}{6(s+3)} \qquad (4.56)$$

and

$$Y(s) = \frac{1}{6(s-3)} - \frac{1}{6(s+3)}. \qquad (4.57)$$

Applying the inverse Laplace transform to (4.56) and (4.57) and using ◼ Table 4.1, we obtain

$$x(t) = \frac{5}{6}\mathcal{L}^{-1}\left\{\frac{1}{s-3}\right\} + \frac{1}{6}\mathcal{L}^{-1}\left\{\frac{1}{s+3}\right\}$$
$$= \frac{5}{6}e^{3t} + \frac{1}{6}e^{-3t}.$$

Similarly, we find

$$y(t) = \frac{1}{6}(e^{3t} - e^{-3t}).$$

Consequently, the solution of (4.54) is

$$(x(t), y(t)) = \frac{1}{6}\left(5e^{3t} + e^{-3t}, e^{3t} - e^{-3t}\right).$$

Exercise 4.2
Use the Laplace transform method to solve the initial value problem

$$y''(t) + 4y'(t) + 4y(t) = t^2, \qquad y(0) = 0, \qquad y'(0) = 0. \qquad (4.58)$$

Solution
Let $Y(s) = \mathcal{L}(y)(s)$. Applying the Laplace transform to both sides in (4.58), we get

$$\mathcal{L}(y''(t) + 4y'(t) + 4y(t)) = \mathcal{L}(t^2).$$

Using the fact that

$$\mathcal{L}(y')(s) = sY(s) - y(0), \qquad \mathcal{L}(y'')(s) = s^2Y(s) - sy(0) - y'(0), \qquad \mathcal{L}(t^2)(s) = \frac{2!}{s^3}$$

together with the linearity of the Laplace transform, we get

$$(s^2 + 4s + 4)Y(s) = \frac{2}{s^3}.$$

Consequently, by solving the above algebraic equation, we obtain

$$Y(s) = \frac{2}{s^3(s+2)^2}.$$

Using the method of the partial fractions, we obtain

$$Y(s) = \frac{2}{s^3(s+2)^2} = \frac{A}{s^3} + \frac{B}{s^2} + \frac{C}{s} + \frac{D}{(s+2)^2} + \frac{E}{s+2}, \qquad (4.59)$$

where A, B, C, D and E are constants. To get A, we multiply both sides in (4.59) by s^3 and take $s = 0$, we obtain $A = 1/2$. To get D, we multiply both sides by $(s + 2)^2$ and take $s = -2$, we get $D = -1/4$. Now, to get the values of the other constants, we multiply (4.59) by $s^3(s + 2)^2$, we obtain

$$2 = A(s + 2)^2 + B(s(s + 2)^2) + C(s^2(s + 2)^2) + Ds^3 + E(s^3(s + 2))$$

By choosing some values of s in the above formula, we can easily get

$$B = -\frac{1}{2}, \qquad C = \frac{3}{8}, \qquad E = -\frac{3}{8}.$$

Now, applying the inverse Laplace transform to (4.59), we obtain

$$y(t) = \mathcal{L}^{-1}(Y(s)) = \mathcal{L}^{-1}\left(\frac{A}{s^3} + \frac{B}{s^2} + \frac{C}{s} + \frac{D}{(s + 2)^2} + \frac{E}{s + 2}\right)$$

$$= \frac{A}{2}t^2 + Bt + C + Dte^{-2t} + Ee^{-2t}.$$

Recalling the values of the above constants, we get the solution of (4.58)

$$y(t) = \frac{1}{4}t^2 - \frac{1}{2}t + \frac{3}{8} - \frac{1}{4}te^{-2t} - \frac{3}{8}e^{-2t}.$$

Exercise 4.3 (Gamma function)

We define the Gamma function as

$$\Gamma(x) = \int_0^\infty u^{x-1}e^{-u}\,du, \qquad u > 0. \tag{4.60}$$

Show that:

1. $\mathcal{L}\{t^x\} = \dfrac{\Gamma(x + 1)}{s^{x+1}}$,
2. $\Gamma(x + 1) = x\Gamma(x)$,
3. $\Gamma(1/2) = \sqrt{\pi}$.

Solution

1. We introduce the change of variables $u = st$, we find $du = s\,dt$, then the integral in (4.60) can be rewritten as

$$\Gamma(x) = \int_0^\infty s(st)^{x-1}e^{-st}\,dt$$

$$= s^x \int_0^\infty t^{x-1}e^{-st}\,dt.$$

Thus, by changing x to $x + 1$, we get

$$\Gamma(x + 1) = s^{x+1}\mathcal{L}\{t^x\},$$

which is the first property in Exercise 4.3.

2. We write $\Gamma(x+1)$ as

$$\Gamma(x+1) = \int_0^\infty u^x e^{-u}\,du.$$

An integration by parts in the above formula yields

$$\Gamma(x+1) = \lim_{N\to\infty} \int_0^N u^x e^{-u}\,du$$

$$= \lim_{N\to\infty}\left(-u^x e^{-u}|_0^N + x\int_0^N u^{x-1}e^{-u}\,du\right)$$

$$= x\int_0^\infty u^{x-1}e^{-u}\,du$$

$$= x\Gamma(x),$$

since $\frac{N^x}{e^N} \to 0$ as $N \to \infty$.

3. To prove the third property in Exercise 4.3, we use the change of variable $u = v^2$, we get

$$\Gamma(1/2) = \int_0^\infty u^{-1/2}e^{-u}\,du$$

$$= 2\int_0^\infty e^{-v^2}\,dv. \tag{4.61}$$

Since the Gauss integral

$$I = \int_{-\infty}^\infty e^{-v^2}\,dv = \sqrt{\pi}$$

and since the function $v \mapsto e^{-v^2}$ is an even function, then we deduce that

$$\Gamma(1/2) = \sqrt{\pi}.$$

Exercise 4.4
Solve the initial value problem in $\mathbb{R} - \{0\}$

$$\begin{cases} y''(t) + 9y(t) = \dfrac{1}{\sqrt{t}}, \\ y(0) = 0, \quad y'(0) = 0. \end{cases} \tag{4.62}$$

Solution
Let $Y(s) = \mathcal{L}\{y\}(s)$. Then, by taking the Laplace transform of both sides in the first equation of (4.62), we get

$$s^2 Y(s) - sy(0) - y'(0) + 9Y(s) = \mathcal{L}\{1/\sqrt{t}\}(s) \tag{4.63}$$

Using the first and the third formulas in Exercise 4.3, we get

$$\mathcal{L}\{1/\sqrt{t}\}(s) = \frac{1}{\sqrt{s}}\Gamma(1/2)$$

$$= \frac{\sqrt{\pi}}{\sqrt{s}}.$$

Plugging this into (4.63) and using the initial values, we get

$$Y(s) = \frac{\sqrt{\pi}}{\sqrt{s}(s^2 + 9)}. \qquad (4.64)$$

Since the denominator in (4.64) does not factor into two polynomials, then we cannot use the partial fractions method to get the inverse Laplace transform. Instead, we use the convolution. Thus, using (4.36), we get

$$y(t) = \mathcal{L}^{-1}\{Y(s)\} = \mathcal{L}^{-1}\left\{\frac{\sqrt{\pi}}{\sqrt{s}}\right\} * \mathcal{L}^{-1}\left\{\frac{1}{s^2 + 9}\right\}$$

$$= \frac{1}{\sqrt{t}} * \frac{1}{3}\sin 3t,$$

where we have used what we have seen above and ◼ Table 4.1. Exploiting (4.32), we find from above that

$$y(t) = \int_0^t \frac{1}{\sqrt{u}}\sin\left(3(t - u)\right)du$$

is the solution of (4.62).

Exercise 4.5 (The unit step function)

Let us consider the unit step function

$$h(t) = \begin{cases} 1 & \text{for } t \geq 0, \\ 0 & \text{for } t < 0. \end{cases}$$

Assume that the Laplace transform $F(s) = \mathcal{L}\{f\}(s)$ of a function $f(t)$ exists and let $a > 0$ be a constant.
1. Show that

$$\mathcal{L}\{h(t - a)f(t - a)\}(s) = e^{-as}F(s) \qquad (4.65)$$

and

$$\mathcal{L}\{h(t - a)f(t)\}(s) = e^{-as}\mathcal{L}\{f(t + a)\}(s). \qquad (4.66)$$

2. Find $\mathcal{L}\{h(t - a)\}(s)$.

Solution

1. Let us first prove (4.65). Indeed, we have by using the definition of the Laplace transform

$$\mathcal{L}\{h(t - a)f(t - a)\}(s) = \int_0^\infty e^{-st}h(t - a)f(t - a)dt.$$

Let us make the change of variable $u = t - a$, then we obtain

$$\mathcal{L}\{h(t-a)f(t-a)\}(s) = \int_{-a}^{\infty} e^{-s(u+a)}h(u)f(u)du$$

$$= e^{-as}\int_{-a}^{\infty} e^{-su}h(u)f(u)du$$

$$= e^{-as}\int_{0}^{\infty} e^{-su}f(u)du$$

$$= e^{-as}F(s),$$

since $h(u) = 0$ for $-a \le u < 0$ and $h(u) = 1$ for $u \ge 0$.
Second, to show (4.66), we can simply apply (4.65) for $g(t) = f(t-a)$ and $g(t+a) = f(t)$, to get

$$\mathcal{L}\{h(t-a)g(t)\}(s) = e^{-as}\mathcal{L}\{g(t+a)\}(s),$$

which has the form (4.66).
2. We need just to apply (4.65) for $f(t) = 1$ to get

$$\mathcal{L}\{h(t-a)\}(s) = e^{-as}\mathcal{L}\{1\}(s)$$

$$= \frac{e^{-as}}{s}.$$

Exercise 4.6
Find the Laplace transform of the function

$$f(t) = \begin{cases} \sin t & \text{if } 0 \le t \le \pi, \\ 0 & \text{if } t > \pi. \end{cases} \tag{4.67}$$

Solution
Using the unit step function in Exercise 4.5, then we may write the function $f(t)$ in (4.67) as

$$f(t) = \sin t - h(t-\pi)\sin t.$$

Taking the Laplace transform of the above equation, we get

$$\mathcal{L}\{f\}(s) = \mathcal{L}\{\sin t - h(t-\pi)\sin t\}(s)$$
$$= \mathcal{L}\{\sin t\}(s) - \mathcal{L}\{h(t-\pi)\sin t\}(s). \tag{4.68}$$

Using ▪ Table 4.1, then the first term on the right-hand side of (4.68) gives

$$\mathcal{L}\{\sin t\}(s) = \frac{1}{s^2+1}. \tag{4.69}$$

On the other hand, exploiting (4.66), we get

$$\mathcal{L}\{h(t-\pi)\sin t\}(s) = e^{-\pi s}\mathcal{L}\{\sin(t+\pi)\}(s)$$
$$= -e^{-\pi s}\mathcal{L}\{\sin t\}(s)$$
$$= -\frac{e^{-\pi s}}{s^2+1}. \tag{4.70}$$

Thus, inserting (4.69) and (4.70) into (4.68), we obtain

$$\mathcal{L}\{f(t)\}(s) = \frac{1+e^{-\pi s}}{s^2+1}.$$

Exercise 4.7 (Duhamel's principle)

We consider the differential equation

$$\begin{cases} ay''(t) + by'(t) + cy(t) = f(t), \\ y(0) = y'(0) = 0. \end{cases} \tag{4.71}$$

where a, b and c are constants and $f(t)$ is an input function such that its Laplace transform $F(s)$ exists. Let $Y(s)$ be the Laplace transform of the output $y(t)$.

1. Find $W(s)$ (usually called the *transfer function*) such that

$$Y(s) = W(s)F(s).$$

2. Show that if $f(t)$ is the unit step function $h(t)$, then

$$Y(s) = Y_h(s) = \frac{W(s)}{s}.$$

3. Let $A(t) = \mathcal{L}^{-1}(Y_h(s))$. Show that the solution of (4.71) is given by

$$y(t) = \int_0^t A'(t) f(t-u) du = \int_0^t A(t-u) f'(u) du + A(t) f(0). \tag{4.72}$$

Formula (4.72) is called the *Duhamel formula*.

Solution

1. Taking the Laplace transform of (4.71), we obtain

$$(as^2 + bs + c)Y(s) = (as + b)y_0 + ay_1 + F(s).$$

This gives, by using the initial values

$$\begin{aligned} Y(s) &= \frac{F(s)}{as^2 + bs + c} \\ &= W(s)F(s) \end{aligned}$$

with

$$W(s) = \frac{1}{as^2 + bs + c}.$$

2. If $f(t) = h(t)$, then

$$\begin{aligned} F(s) &= \int_0^\infty e^{-st} h(t) dt \\ &= \int_0^\infty e^{-st} dt \\ &= \frac{1}{s}, \end{aligned}$$

since $h(t) = 1$ for $t \geq 0$. Consequently,

$$Y_h(s) = \frac{W(s)}{s}. \tag{4.73}$$

3. Now, $Y(s)$ can be written as

$$Y(s) = s\frac{W(s)}{s}F(s)$$
$$= sY_h(s)F(s)$$
$$= sY_h(s)F(s) - A(0)f(0),$$

since $A(0) = 0$. Applying the backward formula of (4.19) and (4.36), we obtain

$$y(t) = \frac{d}{dt}\{A(t) * f(t)\}$$

$$= \frac{d}{dt}\left\{\int_0^t A(w)f(t-w)dw\right\}$$

$$= \frac{d}{dt}\left\{\int_0^t A(t-w)f(w)dw\right\}.$$

Now, by applying the Leibniz rule:

$$\frac{d}{dt}\left\{\int_{\alpha(t)}^{\beta(t)} g(w,t)dw\right\} = \int_{\alpha(t)}^{\beta(t)} \frac{\partial g}{\partial t}(w,t)dw + g(\beta(t),t)\frac{d\beta}{dt}(t) - g(\alpha(t),t)\frac{d\alpha}{dt}(t),$$

where f and $\partial g/\partial t$ are assumed continuous in w and t and $\alpha(t)$ and $\beta(t)$ are differentiable functions of t, we get for $g(t,w) = A(w)f(t-w)$ and $g(t) = A(t-w)f(w)$, $\beta(t) = t$ and $\alpha(t) = 0$,

$$\frac{d}{dt}\left\{\int_0^t A(w)f(t-w)dw\right\} = \int_0^t A(w)f'(t-w)dw + A(t)f(0)$$

and

$$\frac{d}{dt}\left\{\int_0^t A(t-w)f(w)dw\right\} = \int_0^t A'(t-w)f(w)dw.$$

Now, by making the change of variable $u = t - w$, then, we get (4.72).

Exercise 4.8 (Bessel's equation of order zero)
Find the solution (customarily denoted by $J_0(t)$) of the differential equation

$$\begin{cases} ty''(t) + y'(t) + ty(t) = 0, \\ y(0) = 1, \qquad y'(0) = 0. \end{cases} \tag{4.74}$$

Solution
Let us first write

$$\mathcal{L}\{y\}(s) = Y(s), \qquad \mathcal{L}\{y'\}(s) = sY(s) - 1, \qquad \mathcal{L}\{y''\}(s) = s^2Y(s) - s.$$

Next, applying (4.27) for $n = 1$, we get

$$\mathcal{L}\{ty\}(s) = -Y'(s), \quad \mathcal{L}\{ty''\}(s) = -\frac{d}{ds}(s^2Y(s) - s) = -2sY(s) - s^2Y'(s) + 1.$$

Applying the Laplace transform to (4.74) and using the above formulas, we obtain

$$-2sY(s) - s^2Y'(s) + 1 + sY(s) - 1 - Y'(s) = 0.$$

Arranging the above equation, we get

$$(s^2 + 1)Y'(s) + sY(s) = 0. \tag{4.75}$$

Equation (4.75) is a separable first order equation with the dependent variable Y and the independent variable s. We may write (4.75) in the form

$$\frac{Y'(s)}{Y(s)} = -\frac{s}{s^2 + 1}. \tag{4.76}$$

Solving (4.76), by using the method in ▶ Sect. 2.1, we obtain

$$Y(s) = \mathcal{L}\{J_0\}(s) = \frac{C}{\sqrt{s^2 + 1}}$$
$$= \frac{C}{s} \frac{1}{\sqrt{1 + \frac{1}{s^2}}}.$$

Expand this formula with the aid of the binomial series and then compute the inverse Laplace transform to obtain (we leave this for the reader to check)

$$J_0(t) = C \sum_{n=0}^{\infty} \frac{(-1)^n t^{2n}}{2^{2n}(n!)^2}.$$

Finally, since $J_0(0) = 1$, then we obtain $C = 1$.

Exercise 4.9 (The beta function)
We define the beta function as

$$B(m, n) = \int_0^1 t^{m-1}(1 - t)^{n-1} dt, \tag{4.77}$$

where m and n are positive integers. Show that

$$B(m, n) = \frac{\Gamma(m)\Gamma(n)}{\Gamma(m + n)}, \tag{4.78}$$

where $\Gamma(\cdot)$ is the gamma function defined in Exercise 4.3.

Solution
We consider the two functions $f(t) = t^{m-1}$ and $g(t) = t^{n-1}$. Thus, from ◼ Table 4.1, and since for any positive integer p, $\Gamma(p) = (p - 1)!$ we conclude that

$$\mathcal{L}(t^n)(s) = F(s) = \frac{(n-1)!}{s^n}$$
$$= \frac{\Gamma(n)}{s^n}$$

and

$$\mathcal{L}(t^m)(s) = G(s) = \frac{\Gamma(m)}{s^m}.$$

On the other hand, it is clear that form (4.77)

$$B(m, n) = (f * g)(1).$$

Using (4.36), we have

$$(f * g)(t) = \int_0^t u^{m-1}(t-u)^{n-1} du$$

$$= \mathcal{L}^{-1}(F(s)G(s))$$

$$= \Gamma(m)\Gamma(n)\mathcal{L}^{-1}\{s^{-(m+n)}\}$$

$$= \frac{\Gamma(m)\Gamma(n)}{\Gamma(m+n)} t^{m+n-1}.$$

By letting $t = 1$, then (4.78) holds true.

Exercise 4.10

Find the solution of the third order differential equation

$$\begin{cases} y'''(t) + y''(t) = e^t + t + 1, \\ y(0) - y'(0) = y''(0) = 0. \end{cases} \tag{4.79}$$

Solution

Let $Y(s) = \mathcal{L}\{y(t)\}$. Taking the Laplace transform of the first equation in (4.79) and making use of (4.23) and
🔲 Table 4.1, we get

$$(s^3 Y(s) - s^2 y(0) - sy'(0) - y''(0)) + (s^2 Y(s) - sy(0) - y'(0)) = \frac{1}{s-1} + \frac{1}{s^2} + \frac{1}{s}.$$

Using the initial conditions, we obtain

$$s^3 Y(s) + s^2 Y(s) = \frac{2s^2 - 1}{s^2(s-1)}.$$

That is

$$Y(s) = \frac{2s^2 - 1}{s^4(s+1)(s-1)}. \tag{4.80}$$

To find $y(t) = \mathcal{L}^{-1}\{Y(s)\}$, we need to use the method of partial fractions to write the term on the right-hand side of (4.80) as

$$\frac{2s^2 - 1}{s^4(s+1)(s-1)} = \frac{A}{s} + \frac{B}{s^2} + \frac{C}{s^3} + \frac{D}{s^4} + \frac{E}{s+1} + \frac{F}{s-1}.$$

A simple computation leads to $A = C = 0$, $B = -1$, $D = 1$, $E = -1/2$ and $F = 1/2$. Consequently,

$$Y(s) = -\frac{1}{s^2} + \frac{1}{s^4} - \frac{1}{2(s+1)} + \frac{1}{2(s-1)}.$$

Thus, using 🔲 Table 4.1, we obtain

$$y(t) = -\mathcal{L}^{-1}\left\{\frac{1}{s^2}\right\} + \mathcal{L}^{-1}\left\{\frac{1}{s^4}\right\} - \frac{1}{2}\mathcal{L}^{-1}\left\{\frac{1}{s+1}\right\} + \frac{1}{2}\mathcal{L}^{-1}\left\{\frac{1}{s-1}\right\}$$

$$= -t + \frac{1}{6}t^3 - \frac{1}{2}e^{-t} + \frac{1}{2}e^t.$$

Power Series Solution

Belkacem Said-Houari

B. Said-Houari, *Differential Equations: Methods and Applications*, Compact Textbooks in Mathematics,
DOI 10.1007/978-3-319-25735-8_5, © Springer International Publishing Switzerland 2015

5.1 Introduction and Review of Power Series

We have already seen in ▶ Chapter 3 that the solution of differential equations of constant coefficients depends on the solutions of the associated algebraic characteristic equation. There is no similar procedure for solving linear differential equation with variable coefficients. The Laplace transform method that we studied in ▶ Chapter 4 can be an effective tool for some special cases, but as we have seen for the Laplace transform method, it is not always easy to get the inverse Laplace transform as in the case of the solution of the Bessel equation of order n (Exercise 4.8):

$$t^2 y''(t) + t y'(t) + (t^2 - n^2) y(t) = 0.$$

As we will see later, the power series method is a powerful method to solve differential equations with variable coefficients. The basic idea in this method is to assume that the solutions of a given differential equation have power series expansions, and then, we attempt to determine the coefficients in the power series so as to satisfy the differential equation.

In this section, we summarize very briefly (without proof) some theorems that constitute a review of the basic facts about power series. The reader may find the proofs of those theorems in books of calculus.

> **Definition 5.1.1 (Power series)**
>
> A *power series* in $t - t_0$ is an expression of the form
>
> $$\sum_{n=0}^{\infty} a_n (t - t_0)^n = a_0 + a_1(t - t_0) + a_2(t - t_0)^2 + \ldots + a_n(t - t_0)^n + \ldots \qquad (5.1)$$
>
> where t is a variable and the a_n's are constants.

The form (5.1) can be reduced to

$$\sum_{n=0}^{\infty} a_n t^n = a_0 + a_1 t + a_2 t^2 + \ldots + a_n t^n + \ldots \qquad (5.2)$$

by the translation of the coordinate system.

We say that the power series (5.1) converges at the point $t = t_1$ if the infinite series (of real numbers)

$$\sum_{n=0}^{\infty} a_n (t_1 - t_0)^n$$

converges. That is if the sequence of the partial sums

$$S_n = \sum_{k=0}^{n} a_k (t_1 - t_0)^k$$

converges.

Example 5.1

1. It is obvious that (5.1) always converges at the point $t = t_0$ since for $t = t_0$, then

$$\sum_{n=0}^{\infty} a_n (t - t_0)^n = a_0.$$

2. Taking $a_n = 1$ for all $n \geq 0$ in (5.2), then we obtain the geometric power series

$$\sum_{n=0}^{\infty} a_n t^n = 1 + t + t^2 + \ldots + t^n + \ldots \tag{5.3}$$

with first order 1 and ratio t. It converses to $1/(1-t)$ for $|t| < 1$. We express this fact by writing

$$\frac{1}{1-t} = 1 + t + t^2 + \ldots + t^n + \ldots, \qquad -1 < t < 1.$$

3. Now, we take $a_n = 1/n!$ for all n, then (5.2) can be written as

$$\sum_{n=0}^{\infty} \frac{t^n}{n!} = 1 + t + \frac{t^2}{2!} + \ldots + \frac{t^n}{n!} + \ldots \tag{5.4}$$

and converges for all t to e^t. Thus, we write

$$e^t = 1 + t + \frac{t^2}{2!} + \ldots + \frac{t^n}{n!} + \ldots$$

As we will see later, many series have properties similar to (5.3) which means that to each series of this kind we may associate a real number r ($r = 1$ for (5.3)) such that the series converges for $|t| < r$ and diverges for $|t| > r$. This number r is called the *radius of convergence* of the power series.

Theorem 5.1.1 (Radius of convergence)
For each power series $\sum_{n=0}^{\infty} a_n(t - t_0)^n$, there is a number r $(0 \le r \le \infty)$ called the *radius of convergence* of the power series such that the series $\sum_{n=0}^{\infty} a_n(t - t_0)^n$ converges absolutely for $|t - t_0| < r$ and diverges for $|t - t_0| > r$.

If the series converges for all values of t, then $r = \infty$. If the series converges only at $t = t_0$, then $r = 0$.

The power series may or may not converge at either of the end points $t = t_0 - r$ and $t = t_0 + r$.

In many important cases the radius r can be found using the *ratio test* for power series

$$r = \lim_{n\to\infty} \left| \frac{a_n}{a_{n+1}} \right|.$$

Example 5.2
The series

$$\sum_{n=0}^{\infty} \frac{n+1}{3^n} t^n$$

has $r = 3$ as a radius of convergence since

$$\lim_{n\to\infty} \left| \frac{a_n}{a_{n+1}} \right| = \lim_{n\to\infty} \left| \frac{3n+3}{n+2} \right|$$
$$= 3.$$

Definition 5.1.2 (Analytic function)

A function $f(t)$ with the property that a power series expansion of the form

$$f(t) = \sum_{n=0}^{\infty} a_n(t - t_0)^n \qquad (5.5)$$

is valid in some neighborhood of the point t_0 is said to be *analytic* at t_0. In this case the a_n's are necessary given by

$$a_n = \frac{f^{(n)}(t_0)}{n!},$$

and (5.5) is called the *Taylor series* of $f(t)$ at t_0.

Example 5.3
1. Polynomials and e^t function are analytic everywhere.
2. Every rational function is analytic wherever its denominator is nonzero.
3. If the two functions $f(t)$ and $g(t)$ are both analytic at $t = t_0$, then so are their sum $(f + g)(t)$ and their product $(fg)(t)$ and their quotient $(f/g)(t)$ if $g(t_0) \ne 0$.

5.1.1   Power Series Operations

Now, we list (without proof) some operations on the power series.

Theorem 5.1.2 (Identity principle)

If

$$\sum_{n=0}^{\infty} a_n t^n = \sum_{n=0}^{\infty} b_n t^n$$

for every t in some open interval I, then $a_n = b_n$ for all $n \geq 0$.

A theorem from advanced calculus says that a power series can be differentiated term by term at each interior point of its interval of convergence.

Theorem 5.1.3 (Term-by-term differentiation)

Suppose that the series (5.1) converges for $|t - t_0| < r$ with $r > 0$, and denote its sum by

$$f(t) = \sum_{n=0}^{\infty} a_n (t - t_0)^n. \tag{5.6}$$

Then, $f(t)$ is continuous and has derivatives of all orders for $|t - t_0| < r$, and we can obtain the derivatives of $f(t)$ by differentiating the original series (5.6) term by term:

$$f'(t) = \sum_{n=1}^{\infty} n a_n (t - t_0)^{n-1}, \quad f''(t) = \sum_{n=2}^{\infty} n(n-1) a_n (t - t_0)^{n-2},$$

and so on. Each of the resulting series converges for $|t - t_0| < r$.

Example 5.4

Find series for $f'(t)$ and $f''(t)$ if

$$f(t) = \frac{1}{1-t} = \sum_{n=0}^{\infty} t^n = 1 + t + t^2 + \ldots + t^n + \ldots$$

for $-1 < t < 1$.

Solution

Applying Theorem 5.1.3, we obtain for all $-1 < t < 1$:

$$f'(t) = \frac{1}{(1-t)^2} = \sum_{n=1}^{\infty} n t^{n-1} = 1 + 2t + 3t^2 + \ldots + n t^{n-1} + \ldots$$

and

$$f''(t) = \frac{2}{(1-t)^3} = \sum_{n=2}^{\infty} n(n-1)t^{n-2} = 2 + 6t + 12t^2 + \ldots + n(n-1)t^{n-2} + \ldots$$

ℹ Remark 5.1.4 (Shifting the summation index) Just as we make change of variables of integration in a definite integral, it is convenient to make change of summation indices in calculating series solutions in differential equations.

By shifting the index, then we may write (5.1) as

$$\sum_{n=0}^{\infty} a_n(t-t_0)^n = \sum_{n=1}^{\infty} a_{n-1}(t-t_0)^{n-1}$$

$$= \sum_{k=0}^{\infty} a_k(t-t_0)^k,$$

by putting $k = n - 1$.

More generally, we can shift the index of summation by k in (5.1) by simultaneously increasing the summation index k ($n \to n + k$) and decreasing the starting point by k:

$$\sum_{n=k}^{\infty} a_n(t-t_0)^n = \sum_{n=0}^{\infty} a_{n+k}(t-t_0)^{n+k}.$$

If k is negative, we interpret a decrease by k as an increase by $-k$:

$$\sum_{n=0}^{\infty} a_n(t-t_0)^n = \sum_{n=k}^{\infty} a_{n-k}(t-t_0)^{n-k}.$$

Example 5.5

Show that the identity

$$\sum_{n=0}^{\infty} (n+1)a_{n+1}t^n = \sum_{n=1}^{\infty} a_{n-1}t^{n-1} \tag{5.7}$$

implies that

$$a_n = \frac{a_0}{n!}, \qquad n = 1, 2, 3, \ldots$$

Solution

Using Remark 5.1.4, we write (5.7), by shifting the index as

$$\sum_{n=1}^{\infty} n a_n t^{n-1} = \sum_{n=1}^{\infty} a_{n-1}t^{n-1}. \tag{5.8}$$

Applying Theorem 5.1.2 to (5.8), we find

$$a_n = \frac{a_{n-1}}{n!}, \qquad n = 1, 2, 3, \ldots$$

Thus, the above formula gives

$$a_1 = a_0,$$
$$a_2 = \frac{a_1}{2} = \frac{a_0}{2},$$
$$a_3 = \frac{a_2}{3} = \frac{a_0}{3!},$$
$$\vdots$$
$$a_n = \frac{a_0}{n!}$$

and the above formula can be proved easily using the induction on n.

5.2   Series Solutions of Differential Equations

In this section, we show how to use the power series method to find solutions of differential equations. We solve first order and second order differential equations, although the method works also for higher-order differential equations.

5.2.1 Series Solutions of First Order Differential Equations

As we have said before, the method of power series can be used for differential equations that cannot be solved by other methods, especially for differential equations with variable coefficients. To show how the method works, let us take a simple differential equation

$$\begin{cases} y'(t) - y(t) = 0, \\ y(0) = y_0. \end{cases} \tag{5.9}$$

We assume that the initial value problem has a power series solution

$$y(t) = \sum_{n=0}^{\infty} a_n t^n \tag{5.10}$$

that converges for $|t| < r$ with $r > 0$. As we have seen in Theorem 5.1.3, we can differentiate (5.10) term by term to get for $|t| < r$,

$$y'(t) = \sum_{n=1}^{\infty} n a_n t^{n-1}. \tag{5.11}$$

Inserting (5.10) and (5.11) into (5.9), we obtain

$$\sum_{n=1}^{\infty} n a_n t^{n-1} - \sum_{n=0}^{\infty} a_n t^n = 0.$$

By shifting the index in the above identity, we get

$$\sum_{n=0}^{\infty} \Big((n+1) a_{n+1} - a_n \Big) t^n = 0.$$

Hence, Theorem 5.1.2, leads to

$$a_{n+1} = \frac{1}{n+1}a_n, \qquad n = 0, 1, 2, \ldots \tag{5.12}$$

Equation (5.12) is a recurrence relation from which we can compute a_n as

$$a_{n+1} = \frac{1}{(n+1)!}a_0, \qquad n = 0, 1, 2, \ldots \tag{5.13}$$

Plugging (5.13) into (5.10), we find

$$y(t) = y_0 \sum_{n=0}^{\infty} \frac{t^n}{n!}. \tag{5.14}$$

It is clear that the series in (5.14) is the power series expansion of the function e^t. Using the ratio test, we conclude that $r = \infty$. Thus, the solution of (5.9) is

$$y(t) = y_0 e^t.$$

Example 5.6
Find a power series solution of the differential equation

$$(t-3)y'(t) + 2y(t) = 0. \tag{5.15}$$

Solution
As above, we write the solution $y(t)$ as a power series of the form

$$y(t) = \sum_{n=0}^{\infty} a_n t^n.$$

Thus, we have, by differentiating term by term

$$y'(t) = \sum_{n=1}^{\infty} n a_n t^{n-1}.$$

Plugging these into (5.15), we get

$$(t-3) \sum_{n=1}^{\infty} n a_n t^{n-1} + 2 \sum_{n=0}^{\infty} a_n t^n = 0,$$

which can be rewritten as

$$\sum_{n=1}^{\infty} n a_n t^n - 3 \sum_{n=1}^{\infty} n a_n t^{n-1} + 2 \sum_{n=0}^{\infty} a_n t^n = 0.$$

By shifting the index, we find

$$\sum_{n=1}^{\infty} n a_n t^n - 3 \sum_{n=0}^{\infty} (n+1) a_{n+1} t^n + 2 \sum_{n=0}^{\infty} a_n t^n = 0.$$

This leads to

$$2a_0 - 3a_1 + \sum_{n=1}^{\infty} (-3(n+1)a_{n+1} + (2+n)a_n) t^n = 0.$$

Consequently, we obtain from above

$$a_1 = \frac{2}{3}a_0$$

and

$$a_{n+1} = \frac{2+n}{3(n+1)}a_n, \qquad n = 1, 2, 3, \ldots$$

The above two equations lead to

$$a_{n+1} = \frac{2+n}{3(n+1)}a_n, \qquad n = 0, 1, 2, \ldots$$

Thus, we may compute the first terms as

$$a_1 = \frac{2}{3}a_0, \quad a_2 = \frac{3}{3 \cdot 2}a_1 = \frac{3}{3^2}a_0, \quad a_3 = \frac{4}{3^3}a_0.$$

This is enough to make the pattern and we can easily show by induction that

$$a_n = \frac{n+1}{3^n}a_0, \qquad n = 1, 2, \ldots$$

Therefore, our solution $y(t)$ can be written as

$$y(t) = a_0 \sum_{n=0}^{\infty} \frac{n+1}{3^n} t^n.$$

We compute the radius of convergence r as

$$r = \lim_{n\to\infty} \left| \frac{a_n}{a_{n+1}} \right| = \lim_{n\to\infty} \left| \frac{3n+3}{n+2} \right|$$
$$= 3.$$

On the other hand, we have from geometric series

$$\frac{1}{3-t} = \frac{1/3}{1-t/3} = \frac{1}{3} \sum_{n=0}^{\infty} \frac{t^n}{3^n}.$$

Now, differentiating the above series term by term for $|t| < 3$, we get

$$\frac{1}{(3-t)^2} = \frac{1}{3} \sum_{n=1}^{\infty} n \frac{t^{n-1}}{3^n}$$
$$= \frac{1}{3} \sum_{n=0}^{\infty} (n+1) \frac{t^n}{3^{n+1}}$$
$$= \frac{1}{9} \sum_{n=0}^{\infty} (n+1) \frac{t^n}{3^n}.$$

We may choose $y(0) = a_0 = 1/9$, then the solution of (5.15) is

$$y(t) = \frac{1}{(3-t)^2}.$$

5.2.2 Series Solutions of Second Order Differential Equations

The power series method can be applied to differential equations of any order, but its most important applications is to the homogeneous second order equation of the form

$$a(t)y''(t) + b(t)y'(t) + c(t)y(t) = 0, \qquad (5.16)$$

where $a(t) \neq 0$, $b(t)$ and $c(t)$ are analytic functions of t. Equation (5.16) can be written as

$$y''(t) + P(t)y'(t) + Q(t)y(t) = 0, \qquad (5.17)$$

with $P(t) = b(t)/a(t)$ and $Q(t) = c(t)/a(t)$.

Definition 5.2.1 (Ordinary and singular points)

A point t_0 is called an *ordinary point* of equation (5.17) if both $P(t)$ and $Q(t)$ are analytic at t_0. If t_0 is not an ordinary point, it is called a *singular point* of the equation.

Example 5.7

The differential equation

$$y''(t) - \frac{2t}{1-t^2}y' + \frac{2}{1-t^2}y = 0$$

has two singular points $t = 1$ and $t = -1$ since the coefficients $-\frac{2t}{1-t^2}$ and $\frac{2}{1-t^2}$ are not analytic at these two points.

The following theorem justifies the power series method and gives the nature of solutions near ordinary points.

Theorem 5.2.1

Let t_0 be an ordinary point of the equation (5.17) and let a_0 and a_1 be arbitrary constants. Then there exists a unique function $y(t)$ that is analytic at t_0, is a solution of equation (5.17) in a certain neighborhood of this point and satisfies the initial conditions $y(t_0) = a_0$ and $y'(t_0) = a_1$. Furthermore, if the power series expansions of $P(t)$ and $Q(t)$ are valid on an interval $|t - t_0| < r$, $r > 0$, then the power series expansion of this solution is also valid in the same interval.

We omit the proof of the above theorem since it is technical and can be found in many textbooks.

Example 5.8

Find the solution of the initial value problem

$$\begin{cases} y''(t) + y(t) = 0, \\ y(0) = y_0, \qquad y'(0) = y_1. \end{cases} \qquad (5.18)$$

Solution

It is clear that $t_0 = 0$ is an ordinary point of (5.18). Thus, we look for the solution in the form of power series about $t_0 = 0$. Indeed, let

$$y(t) = \sum_{n=0}^{\infty} a_n t^n$$

be the power series expansion of the solution of (5.18). Thus, differentiating term by term, we obtain

$$y'(t) = \sum_{n=1}^{\infty} n a_n t^{n-1}, \qquad y''(t) = \sum_{n=2}^{\infty} n(n-1) a_n t^{n-2}.$$

Consequently, plugging the above formulas into (5.18), we obtain

$$\sum_{n=0}^{\infty} a_n t^n + \sum_{n=2}^{\infty} n(n-1) a_n t^{n-2} = 0.$$

By shifting the index in the above identity, we obtain

$$\sum_{n=0}^{\infty} \left((n+2)(n+1) a_{n+2} + a_n \right) t^n = 0.$$

This leads to

$$a_{n+2} = -\frac{1}{(n+2)(n+1)} a_n, \qquad n = 0, 1, 2, \ldots$$

The above formula gives

$$a_2 = -\frac{1}{1 \cdot 2} a_0, \qquad a_3 = -\frac{1}{2 \cdot 3} a_1.$$

Thus, we can find the pattern as follows:

$$a_{2k} = \frac{(-1)^k}{(2k)!} a_0, \qquad a_{2k+1} = \frac{(-1)^k}{(2k+1)!} a_1, \qquad k = 1, 2, \ldots$$

We may easily prove the above formulas using the induction. Thus, our solution is written as

$$y(t) = a_0 \sum_{k=0}^{\infty} \frac{(-1)^k}{(2k)!} t^{2k} + a_1 \sum_{k=0}^{\infty} \frac{(-1)^k}{(2k+1)!} t^{2k+1}. \tag{5.19}$$

Using the ratio test, we see that in both series $r = \infty$. It is clear that the first series in (5.19) is the power series expansion of $\cos t$ and the second one is the power series expansion of $\sin t$. Thus,

$$y(t) = a_0 \cos t + a_1 \sin t.$$

Using the initial values, we find that $a_0 = y_0$ and $a_1 = y_1$. Then, the solution of (5.18) is

$$y(t) = y_0 \cos t + y_1 \sin t.$$

5.3   Exercises

Exercise 5.1 (Lagender's equation)

Find the general solution of the Lagender differential equation

$$(1 - t^2)y''(t) - 2ty'(t) + p(p + 1)y(t) = 0 \qquad (5.20)$$

in terms of power series in t, where p is a constant.

Solution

Equation (5.20) can be written as

$$y''(t) - \frac{2t}{1 - t^2}y'(t) + \frac{p(p + 1)}{1 - t^2}y(t) = 0. \qquad (5.21)$$

Since

$$P(t) = -\frac{2t}{1 - t^2} \qquad \text{and} \qquad Q(t) = \frac{p(p + 1)}{1 - t^2}$$

are analytic functions at $t = 0$, then $t = 0$ is an ordinary point. Thus, we look for a solution of (5.20) written in power series expansion of the form

$$y(t) = \sum_{n=0}^{\infty} a_n t^n.$$

Differentiating term by term, we find

$$y'(t) = \sum_{n=1}^{\infty} n a_n t^{n-1}, \qquad y''(t) = \sum_{n=2}^{\infty} n(n - 1) a_n t^{n-2}.$$

Inserting the above identities into (5.20), we get

$$(1 - t^2)\sum_{n=2}^{\infty} n(n - 1) a_n t^{n-2} - 2t\sum_{n=1}^{\infty} n a_n t^{n-1} + p(p + 1)\sum_{n=0}^{\infty} a_n t^n = 0. \qquad (5.22)$$

Multiplying by the coefficients and shifting the index, we obtain

$$\sum_{n=0}^{\infty}(n + 2)(n + 1) a_{n+2} t^n - \sum_{n=2}^{\infty} n(n - 1) a_n t^n - \sum_{n=1}^{\infty} 2n a_n t^n + \sum_{n=0}^{\infty} p(p + 1) a_n t^n = 0.$$

This leads to

$$2a_2 + p(p + 1)a_0 + (6a_3 + (p(p + 1) - 2)a_1)t$$

$$+ \sum_{n=2}^{\infty} \left\{ (n + 2)(n + 1)a_{n+2} + (p(p + 1) - n(n + 1)) \right\} t^n = 0.$$

Applying Theorem 5.1.2, we obtain

$$\begin{cases} a_2 = -\dfrac{p(p + 1)}{2} a_0, \\ a_3 = -\dfrac{p(p + 1) - 2}{6} a_1, \\ a_{n+2} = -\dfrac{p(p + 1) - n(n + 1)}{(n + 2)(n + 1)} a_n, \quad n = 2, 3, \dots \end{cases} \qquad (5.23)$$

The last formula in (5.23) can be also written as

$$a_{n+2} = -\frac{(p-n)(p+n+1)}{(n+2)(n+1)}a_n, \quad n = 2, 3, \ldots$$

In view of the first and the second formulas in (5.23), we get

$$a_{n+2} = -\frac{(p-n)(p+n+1)}{(n+2)(n+1)}a_n, \quad n = 0, 1, 2, \ldots \tag{5.24}$$

Now, let us assume that $p > -1$. If $p \leq -1$, then, we put $p = -(1+\gamma)$, $\gamma > 0$, which leads to the Lagender equation

$$(1-t^2)y''(t) - 2ty'(t) + \gamma(\gamma+1)y(t) = 0.$$

To find the pattern, we may compute the first terms as

$$\begin{cases} a_2 = -\dfrac{p(p+1)}{1 \cdot 2}a_0, \\ a_3 = -\dfrac{p(p+1)-2}{1 \cdot 2 \cdot 3}a_1, \\ a_4 = -\dfrac{(p-1)(p+3)}{3 \cdot 4}a_2 = \dfrac{p(p-2)(p+1)(p+3)}{4!}a_0, \\ a_5 = -\dfrac{(p-3)(p+4)}{4 \cdot 5}a_3 = \dfrac{(p-1)(p-3)(p+2)(p+4)}{5!}a_1. \end{cases}$$

Thus, we may deduce that

$$\begin{cases} a_{2n} = (-1)^n \dfrac{p(p-2)(p-4)\cdots(p-2n+2)(p+1)(p+3)\cdots(p+2n-1)}{(2n)!} \\ a_{2n+1} = (-1)^n \dfrac{(p-1)(p-3)\cdots(p-2n-1)(p+2)(p+4)\cdots(p+2n)}{(2n+1)!} \end{cases}$$

Consequently, we get two power series linearly independent and since $a_0 = y(0) = y_0$ and $a_1 = y'(0) = y_1$, then the solution of (5.20) is given by

$$y(t) = y_0 \sum_{n=0}^{\infty} a_{2n}t^{2n} + y_1 \sum_{n=0}^{\infty} a_{2n+1}t^{2n+1}.$$

It is clear that if p is not an integer, then the radius of convergence of both series in the above formula is $r = 1$. This can be easily seen by using the ratio test formula.

Exercise 5.2 (Binomial series)
Find the power series expansion of the function

$$y(t) = (1+t)^p, \tag{5.25}$$

where p is an arbitrary constant.

Solution
It is not difficult to see that $y(t)$ is the solution to the initial value problem

$$\begin{cases} (1+t)y'(t) - py(t) = 0, \\ y(0) = 1. \end{cases} \tag{5.26}$$

We assume that (5.25) has a power series expansion about $t = 0$ with positive radius of convergence. Thus, we write $y(t)$ as

$$y(t) = \sum_{n=0}^{\infty} a_n t^n.$$

Differentiating term by term we get

$$y'(t) = \sum_{n=1}^{\infty} n a_n t^{n-1}.$$

Plugging these into (5.26), we get

$$(1 + t) \sum_{n=1}^{\infty} n a_n t^{n-1} - p \sum_{n=0}^{\infty} a_n t^n = 0.$$

This leads to

$$\sum_{n=1}^{\infty} n a_n t^{n-1} + \sum_{n=1}^{\infty} n a_n t^n - p \sum_{n=0}^{\infty} a_n t^n = 0.$$

Shifting the index in the above formula, we obtain

$$(a_1 - p a_0) + \sum_{n=1}^{\infty} \left\{ (n+1) a_{n+1} - a_n (p - n) \right\} t^n = 0.$$

Consequently, Theorem 5.1.2 implies

$$\begin{cases} a_1 = p a_0 \\ a_{n+1} = \dfrac{p - n}{n + 1} a_n, & n = 1, 2, \ldots \end{cases}$$

This can be written as

$$a_{n+1} = \frac{p - n}{n + 1} a_n, \qquad n = 0, 1, 2, \ldots \tag{5.27}$$

Since $y(0) = a_0 = 1$, then (5.27) gives

$$\begin{cases} a_1 = p, \qquad a_2 = \dfrac{p - 1}{2} a_1 = \dfrac{p(p - 1)}{2}, \\ a_3 = \dfrac{p - 2}{3} a_2 = \dfrac{p(p - 1)(p - 2)}{2 \cdot 3}, \\ \vdots \\ a_n = \dfrac{p(p - 1)(p - 2) \cdots (p - n + 1)}{n!}. \end{cases}$$

Thus, the power series expansion of (5.25) is given by

$$y(t) = 1 + p t + \frac{p(p - 1)}{2!} t^2 + \ldots + \frac{p(p - 1)(p - 2) \cdots (p - n + 1)}{n!} t^n + \ldots \tag{5.28}$$

We may find the radius of convergence r as

$$\lim_{n \to \infty} \left| \frac{a_n}{a_{n+1}} \right| = \lim_{n \to \infty} \left| \frac{(p - n + 2) n!}{(n + 1)!} \right| = 1.$$

Thus, the series (5.28) converges for $|t| < 1$ and diverges for $|t| > 1$.

Exercise 5.3 (Airy's equation)

Find the power series solution of the Airy equation

$$y''(t) - ty(t) = 0. \tag{5.29}$$

Solution

We search for a solution of the form

$$y(t) = \sum_{n=0}^{\infty} a_n t^n. \tag{5.30}$$

Taking the second derivative in (5.30) and plugging the result into (5.29), we obtain

$$\sum_{n=2}^{\infty} n(n-1)a_n t^{n-2} - \sum_{n=0}^{\infty} a_n t^{n+1} = 0.$$

By shifting the index in the above formula, we get

$$\sum_{n=0}^{\infty} (n+2)(n+1)a_{n+2} t^n - \sum_{n=1}^{\infty} a_{n-1} t^n = 0.$$

This can be rewritten as

$$2a_2 + \sum_{n=1}^{\infty} \left((n+2)(n+1)a_{n+2} - a_{n-1} \right) t^n = 0.$$

Applying Theorem 5.1.2, we get

$$\begin{cases} a_2 = 0, \\ a_{n+2} = \dfrac{a_{n-1}}{(n+2)(n+1)}, \quad n \geq 1. \end{cases}$$

From the above formula, we may also write

$$a_m = \frac{1}{m(m-1)} a_{m-3}, \quad m \geq 3.$$

Now, if $a_0 = 1$ and $a_1 = 0$, we deduce that

$$a_2 = a_5 = a_8 = \ldots = 0, \quad a_1 = a_4 = a_7 = \ldots = 0$$

and

$$a_3 = \frac{1}{2 \cdot 3}, \quad a_6 = \frac{1}{5 \cdot 6} a_3 = \frac{1}{(5 \cdot 6)(2 \cdot 3)}.$$

Thus, in this case we can easily find the pattern as

$$a_{3k} = \frac{1}{3k(3k-1)} a_{3(k-1)} = \frac{1}{3k(3k-1)\cdots(5 \cdot 6)(2 \cdot 3)}$$
$$= \frac{4 \cdot 7 \cdots (3k-2)}{(3k)!}, \quad k \geq 1.$$

This gives the first solution of (5.29) as

$$y_1(t) = 1 + \sum_{k=1}^{\infty} \frac{1 \cdot 4 \cdot 7 \cdots (3k-2)}{(3k)!} t^{3k}.$$

Similarly, if $a_0 = 0$ and $a_1 = 1$, then we have

$$a_2 = a_5 = a_8 = \ldots = 0, \qquad a_0 = a_3 = a_6 = \ldots = 0$$

and

$$a_{3k+1} = \frac{1}{(3k+1)3k \cdots (6 \cdot 7)(3 \cdot 4)} = \frac{2 \cdot 5 \cdot 8 \cdots (3k-1)}{(3k+1)!}, \qquad k \geq 1.$$

This gives

$$y_2(t) = t + \sum_{k=1}^{\infty} \frac{2 \cdot 5 \cdot 8 \cdots (3k-1)}{(3k+1)!} t^{3k+1}.$$

We may easily show that the radius of convergence is $r = \infty$. Thus, the solutions $y_1(t)$ and $y_2(t)$ are two independent solutions of (5.29).

Exercise 5.4
We consider the differential equation

$$2ty'(t) + y(t) - 3t \cos(t^{3/2}) = 0. \tag{5.31}$$

Find the power series solution of (5.31) defined on $(0, \infty)$ and the sum of the series solution.

Solution
Let

$$y(t) = \sum_{n=0}^{\infty} a_n t^n,$$

where a_n is a real number for all n in $\mathbb{N}$, be the solution written in power series, of (5.31). We have

$$y'(t) = \sum_{n=1}^{\infty} n a_n t^{n-1}.$$

On the other hand, we have for all t in $(0, \infty)$,

$$\cos(t^{3/2}) = \sum_{n=0}^{\infty} \frac{(-1)^n t^{3n}}{(2n)!}.$$

Thus, we get

$$3t \cos(t^{3/2}) = \sum_{n=0}^{\infty} \frac{3(-1)^n t^{3n+1}}{(2n)!}.$$

Consequently, from above, we get

$$2ty'(t) + y(t) - 3t\cos(t^{3/2}) = \sum_{n=1}^{\infty} 2na_n t^n + \sum_{n=0}^{\infty} a_n t^n - \sum_{n=0}^{\infty} \frac{3(-1)^n t^{3n+1}}{(2n)!}$$

$$= 0.$$

Shifting the index in the above formula, we get

$$a_0 + \sum_{n=1}^{\infty} (2n+1)a_n t^n = \sum_{n=0}^{\infty} \frac{3(-1)^n t^{3n+1}}{(2n)!}.$$

It is clear form the above identity that

$$\begin{cases} a_n = 0, & \text{if } n = 3k, \text{ or } n = 3k + 2, \\ a_n = \dfrac{(-1)^k}{(2k+1)!}, & \text{if } n = 3k + 1 \text{ and } k \geq 0. \end{cases}$$

Thus, the solution of (5.31) is

$$y(t) = \sum_{k=0}^{\infty} \frac{(-1)^k}{(2k+1)!} t^{3k+1}. \tag{5.32}$$

Using the ratio test, we find that the radius of convergence of the series in (5.32) is $r = \infty$.
On the other hand, we have

$$\sin t = \sum_{k=0}^{\infty} (-1)^k \frac{t^{2k+1}}{(2k+1)!}.$$

Thus, it is clear that for t in $(0, \infty)$, then we have

$$\sin(t^{3/2}) = \sum_{k=0}^{\infty} (-1)^k \frac{t^{3k+3/2}}{(2k+1)!}.$$

Hence,

$$\frac{\sin(t^{3/2})}{\sqrt{t}} = \sum_{k=0}^{\infty} (-1)^k \frac{t^{3k+1}}{(2k+1)!}. \tag{5.33}$$

Comparing (5.32) and (5.33), we deduce that the sum of the series solution of (5.31), is given by

$$y(t) = \frac{\sin(t^{3/2})}{\sqrt{t}}.$$

Systems of Differential Equations

Belkacem Said-Houari

B. Said-Houari, *Differential Equations: Methods and Applications*, Compact Textbooks in Mathematics, DOI 10.1007/978-3-319-25735-8_6, © Springer International Publishing Switzerland 2015

6.1 Introduction

Systems of differential equations arise in many scientific problems, for instance we have seen in ▶ Sect. 1.1.3 that the predator–prey interaction can be modeled by the system of differential equations (1.10).

Let $y_1(t)$, $y_2(t), \ldots, y_n(t)$ are unknown functions of a single independent variable t, the most interesting systems in applications are systems of the form

$$\begin{cases} y_1'(t) = f_1(t, y_1(t), y_2(t), \ldots, y_n(t)), \\ y_2'(t) = f_2(t, y_1(t), y_2(t), \ldots, y_n(t)), \\ \vdots \\ y_n'(t) = f_n(t, y_1(t), y_2(t), \ldots, y_n(t)), \end{cases} \tag{6.1}$$

where we have n dependent variables $y_1, y_2, \ldots, y_n$ and one independent variable t. Later on, we may drop the t in order to shorten the notation.

It is not hard to see that any single differential equation of nth order of the form

$$y^{(n)} = f\left(t, y, y', y'', \ldots, y^{(n-1)}\right) \tag{6.2}$$

can be written as a system of first order differential equations of the form (6.1). Indeed, we introduce the new variables

$$y_1 = y, \quad y_2 = y', \quad \ldots, \quad y_n = y^{(n-1)}, \tag{6.3}$$

then we obtain

$$\begin{cases} y_1' = y_2, \\ y_2' = y_3, \\ \vdots \\ y_n' = f(t, y_1, y_2, \ldots, y_n). \end{cases} \tag{6.4}$$

More detailed discussion on the linear version of (6.2) is given in ▶ Sect. 6.2.7.

As we will see later, it is much easier to deal with system (6.4) rather than equation (6.2).

Example 6.1

The second order differential equation studied in ▶ Chapter 3

$$a(t)y'' + b(t)y' + c(t)y = 0, \qquad a \neq 0, \tag{6.5}$$

can be written as a first order system of two dependent variables y_1 and y_2 and one independent variable t as follows: we put

$$y_1 = y, \quad \text{and} \quad y_2 = y',$$

then we obtain

$$\begin{cases} y_1' = y_2, \\ y_2' = -\dfrac{c(t)}{a(t)} y_1 - \dfrac{b(t)}{a(t)} y_2. \end{cases} \tag{6.6}$$

Definition 6.1.1 (Linear first order system)

If the functions f_i, $i = 1, \ldots, n$ in (6.1) have the form

$$f_i(t, y_1, y_2, \ldots, y_n) = a_{i1}(t)y_1 + a_{i2}(t)y_2 + \ldots + a_{in}(t)y_n + b_i(t), \quad i = 1, \ldots, n, \tag{6.7}$$

where $a_{ij}(t)$, $1 \leq i, j \leq n$ and $b_i(t)$, $1 \leq i \leq n$ are functions depending only on t, then system (6.1) is called a *first-order linear system* of differential equations.

It is clear that system (6.6) is a linear system.

We assume that the functions $a_{ij}(t)$, $1 \leq i, j \leq n$, $b_i(t)$, $1 \leq i \leq n$ are continuous on a certain interval J.

If system (6.1) is linear, then we can rewrite it using the matrix from

$$\frac{d}{dt} \begin{bmatrix} y_1 \\ y_2 \\ \vdots \\ y_n \end{bmatrix} = \begin{bmatrix} a_{11} & a_{12} & \cdots & a_{1n} \\ a_{21} & a_{22} & \cdots & a_{2n} \\ \vdots & & & \vdots \\ a_{n1} & a_{n2} & \cdots & a_{nn} \end{bmatrix} \begin{bmatrix} y_1 \\ y_2 \\ \vdots \\ y_n \end{bmatrix} + \begin{bmatrix} b_1 \\ b_2 \\ \vdots \\ b_n \end{bmatrix}. \tag{6.8}$$

Or equivalently

$$\frac{d}{dt} Y(t) = A(t)Y(t) + B(t), \tag{6.9}$$

where $Y(t)$ and $B(t)$ are the vectors

$$Y(t) = \begin{bmatrix} y_1(t) \\ y_2(t) \\ \vdots \\ y_n(t) \end{bmatrix}, \qquad B(t) = \begin{bmatrix} b_1(t) \\ b_2(t) \\ \vdots \\ b_n(t) \end{bmatrix} \tag{6.10}$$

and $A(t)$ is the square matrix

$$A(t) = \begin{bmatrix} a_{11}(t) & a_{12}(t) & \cdots & a_{1n}(t) \\ a_{21}(t) & a_{22}(t) & \cdots & a_{2n}(t) \\ \vdots & \vdots & & \vdots \\ a_{n1}(t) & a_{n2}(t) & \cdots & a_{nn}(t) \end{bmatrix}. \tag{6.11}$$

Example 6.2

Express the system of differential equation

$$\begin{cases} \dfrac{dy_1}{dt} = y_1 \sin t - y_2 \cos t + t, \\ \dfrac{dy_2}{dt} = y_1 \cos t + y_2 \sin t + 3, \end{cases} \tag{6.12}$$

in its matrix form.

Solution

We may easily write (6.12) in the form (6.8) as

$$\frac{d}{dt}\begin{bmatrix} y_1 \\ y_2 \end{bmatrix} = \begin{bmatrix} \sin t & -\cos t \\ \cos t & \sin t \end{bmatrix}\begin{bmatrix} y_1 \\ y_2 \end{bmatrix} + \begin{bmatrix} t \\ 3 \end{bmatrix}. \tag{6.13}$$

If $B(t) = 0$, that is $b_i(t) = 0$, $1 \le i \le n$, then system (6.9) is called *homogeneous*; otherwise it is said to be *nonhomogeneous*.

Example 6.3

Find the solution of the linear homogeneous system

$$\begin{cases} \dfrac{dx}{dt} = 4x - y, \\ \dfrac{dy}{dt} = 2x + y. \end{cases} \tag{6.14}$$

Solution

Using the matrix form, then system (6.14) can be written as

$$\frac{d}{dt}\begin{bmatrix} x(t) \\ y(t) \end{bmatrix} = \begin{bmatrix} 4 & -1 \\ 2 & 1 \end{bmatrix}\begin{bmatrix} x(t) \\ y(t) \end{bmatrix}. \tag{6.15}$$

By taking the derivative of the second equation in (6.14), we get

$$y'' = 2x' + y'. \tag{6.16}$$

Multiplying the second equation in (6.14) by -2 and add the result to the first one, we get

$$x' - 2y' = -3y. \tag{6.17}$$

Inserting equation (6.17) into (6.16), we obtain

$$y'' - 5y' + 6y = 0. \tag{6.18}$$

Using the method discussed in ▶ Sect. 3.1.2, we find that

$$y_1(t) = e^{2t} \quad \text{and} \quad y_2(t) = e^{3t} \tag{6.19}$$

are two linearly independent solutions of (6.18). Thus, the solution of (6.18) is

$$y(t) = c_1 e^{2t} + c_2 e^{3t}.$$

The second equation in (6.14) gives

$$
\begin{aligned}
x(t) &= \frac{1}{2}(y'(t) - y(t)) \\
&= \frac{c_1}{2} e^{2t} + c_2 e^{3t}.
\end{aligned}
$$

Consequently, the solution of system (6.14) is

$$
\begin{cases}
x(t) = \dfrac{c_1}{2} e^{2t} + c_2 e^{3t}, \\
y(t) = c_1 e^{2t} + c_2 e^{3t}.
\end{cases}
$$

This can be written using the vector notation as

$$
\begin{aligned}
\begin{bmatrix} x(t) \\ y(t) \end{bmatrix} &= \frac{c_1}{2} \begin{bmatrix} 1 \\ 2 \end{bmatrix} e^{2t} + c_2 \begin{bmatrix} 1 \\ 1 \end{bmatrix} e^{3t} \\
&= \frac{c_1}{2} V_1 e^{2t} + c_2 V_2 e^{3t},
\end{aligned} \tag{6.20}
$$

where V_1 and V_2 are the two vectors

$$
V_1 = \begin{bmatrix} 1 \\ 2 \end{bmatrix}, \qquad V_2 = \begin{bmatrix} 1 \\ 1 \end{bmatrix}.
$$

6.2   First Order Linear Systems

In order to introduce the method of solving linear systems written in the matrix form, let us try to solve the system (6.14) given in Example 6.3 directly and without using its equivalent second order differential equation (6.18). Indeed, as we have seen in ▶ Sect. 3.1.2, we may try first to look for a solution of the form

$$
\begin{bmatrix} x(t) \\ y(t) \end{bmatrix} = \begin{bmatrix} a_1 \\ a_2 \end{bmatrix} e^{\lambda t}.
$$

where a_1, a_2 and λ are constants. Thus, taking the derivative of the above formula and plugging it into (6.15), we obtain

$$\lambda \begin{bmatrix} a_1 \\ a_2 \end{bmatrix} e^{\lambda t} = \begin{bmatrix} 4 & -1 \\ 2 & 1 \end{bmatrix} \begin{bmatrix} a_1 \\ a_2 \end{bmatrix} e^{\lambda t}. \tag{6.21}$$

This leads to the system of algebraic equations

$$\begin{cases} \lambda a_1 = 4a_1 - a_2, \\ \lambda a_2 = 2a_1 + a_2. \end{cases} \tag{6.22}$$

If we look for λ as just a parameter, then system (6.22) is a linear system with the unknowns a_1 and a_2 and can be rewritten as

$$\begin{cases} (4 - \lambda)a_1 - a_2 = 0, \\ 2a_1 + (1 - \lambda)a_2 = 0. \end{cases} \tag{6.23}$$

The only possibility for (6.23) not having a trivial solution $a_1 = a_2 = 0$ is if the determinant of (6.23) is identically zero. That is if

$$\det \begin{bmatrix} 4 - \lambda & -1 \\ 2 & 1 - \lambda \end{bmatrix} = \det(A - \lambda I) = 0, \tag{6.24}$$

where A and I are the matrices

$$A = \begin{bmatrix} 4 & -1 \\ 2 & 1 \end{bmatrix}, \qquad I = \begin{bmatrix} 1 & 0 \\ 0 & 1 \end{bmatrix}.$$

That is if λ is an eigenvalue (see Definition 6.2.5) of the matrix A. Equation (6.24) leads to the equation

$$\lambda^2 - 5\lambda + 6 = 0,$$

which is the characteristic equation associated to (6.18). Solving the above equation, we find that $\lambda = 3$ and $\lambda = 2$ are the two solutions. Thus, for $\lambda = 3$, then system (6.23) leads to just one equation

$$a_1 - a_2 = 0.$$

That is

$$a_2 = a_1.$$

Similarly, for $\lambda = 2$, we obtain the equation

$$2a_1 = a_2.$$

Consequently, the vector

$$V_2 = \begin{bmatrix} 1 \\ 1 \end{bmatrix}$$

is the eigenvector associated to the eigenvalue $\lambda = 3$ and

$$V_1 = \begin{bmatrix} 1 \\ 2 \end{bmatrix}$$

is the eigenvector associated to the eigenvalue $\lambda = 2$.
Therefore, we obtain two solutions

$$\begin{bmatrix} x(t) \\ y(t) \end{bmatrix} = a_1 \begin{bmatrix} 1 \\ 1 \end{bmatrix} e^{3t} \quad \text{and} \quad \begin{bmatrix} x(t) \\ y(t) \end{bmatrix} = a_1 \begin{bmatrix} 1 \\ 2 \end{bmatrix} e^{2t}.$$

Since the system (6.14) is linear, then according to the superposition principle (Theorem 3.3.6) any linear combination of the above two solutions is also a solution to (6.14). That is

$$\begin{bmatrix} x(t) \\ y(t) \end{bmatrix} = C_1 V_1 e^{2t} + C_2 V_2 e^{3t},$$

which is exactly the solution given in (6.20) since C_1 and C_2 are arbitrary constants.
We may easily generalize the above procedure to the homogeneous system

$$\frac{d}{dt} Y(t) = AY(t), \tag{6.25}$$

where $Y(t)$ is the vector defined in (6.10) and A is the matrix in (6.11) such that $a_{ij}(t)$, $1 \le i, j \le n$ are constants. Thus, as before, we may look for a solution of the form

$$Y(t) = Ve^{\lambda t}$$

where

$$V = \begin{bmatrix} a_1 \\ a_2 \\ \vdots \\ a_n \end{bmatrix}$$

is a vector of n components. Hence, we may easily see as above that for $Y(t)$ to be a solution of (6.25), then λ should satisfy the characteristic equation

$$\det(A - \lambda I) = 0$$

where I is the $n \times n$ identity matrix. We will discuss later in ▶ Sect. 6.2.4, the general method of solving linear first order systems with constant coefficients.

ⓘ Remark 6.2.1 If A is 2×2 matrix, then the characteristic equation is

$$\det(A - \lambda I) = \lambda^2 - \operatorname{tr}(A)\lambda + \det(A), \tag{6.26}$$

where $\operatorname{tr}(A)$ denotes the trace of A.

Let us start with the following definition which extends Definition 3.1.1 to the vector valued functions.

Definition 6.2.1 (Linear dependence of vector functions)

Let $Y_1, Y_2, \ldots, Y_n$ be n vector functions such that

$$Y_j = \begin{bmatrix} y_{1j} \\ y_{2j} \\ \vdots \\ y_{nj} \end{bmatrix}, \quad 1 \le j \le n. \tag{6.27}$$

These vector functions are said to be *linearly dependent* on some interval J if there exist constants $c_1, c_2, \ldots, c_n$ not all zero such that

$$c_1 Y_1(t) + c_2 Y_2(t) + \ldots + c_n Y_n(t) = 0, \tag{6.28}$$

for all t in J. If the vectors are not linearly dependent, they are said to be *linearly independent* on J.

Example 6.4

The vectors

$$Y_1(t) = \begin{bmatrix} e^{3t} \\ e^{3t} \end{bmatrix}, \quad \text{and} \quad Y_2(t) = \begin{bmatrix} e^{2t} \\ 2e^{2t} \end{bmatrix}$$

found in Example 6.14 are linearly independent on $\mathbb{R}$, since if there are two constants c_1 and c_2 such that

$$c_1 Y_1(t) + c_2 Y_2(t) = 0,$$

then $c_1 = c_2 = 0$. That is (6.28) is not satisfied.

Now, we extend the definition of the Wronskian given in Definition 3.1.2 to vector functions as follows.

Definition 6.2.2 (Wronskian of vector functions)

The Wronskian of the vector functions Y_j, $1 \leq j \leq n$ in (6.27) is defined to be the real valued function

$$W[Y_1, Y_2, \ldots, Y_n](t) = \det \begin{bmatrix} y_{11}(t) & y_{12}(t) & \cdots & y_{1n}(t) \\ y_{21}(t) & y_{22}(t) & \cdots & y_{2n}(t) \\ \vdots & \vdots & & \vdots \\ y_{n1}(t) & y_{n2}(t) & \cdots & y_{nn}(t) \end{bmatrix}.$$

Now, we may extend Lemma 3.1.1 to a system of differential equations as follows:

Lemma 6.2.2 Suppose that $Y_1, Y_2, \ldots, Y_n$ are n solutions to the linear homogeneous system (6.25) on an open interval J. Assume also that $A(t)$ is continuous on J and let

$$W(t) = W[Y_1, Y_2, \ldots, Y_n](t).$$

Then

- If $Y_1, Y_2, \ldots, Y_n$ are linearly dependent on J, then $W(t) = 0$ for all t in J.
- If $Y_1, Y_2, \ldots, Y_n$ are linearly independent on J, then $W(t) \neq 0$ for all t in J.

Thus, there are only two possibilities for solution of the homogeneous system (6.25): either $W(t) = 0$ at every point of J, or $W(t) = 0$ at no point of J.

The proof of Lemma 6.2.2 can be done with some slight modification of the one of Lemma 3.1.1. We omit it.

Example 6.5
If we take the vectors $Y_1(t)$ and $Y_2(t)$ defined in Example 6.4, we find that the Wronskian of these vectors is

$$W(t) = W[Y_1, Y_2](t) = \det \begin{bmatrix} e^{3t} & e^{2t} \\ e^{3t} & 2e^{2t} \end{bmatrix} = e^{5t}$$

which is never zero. Thus the vectors $Y_1(t)$ and $Y_2(t)$ are linearly independent on any open interval.

Theorem 6.2.3 (General solution of the homogeneous system)

Suppose that $Y_1, Y_2, \ldots, Y_n$ are n solutions linearly independent to the linear homogeneous system (6.25) on an open interval J, where $A(t)$ is continuous. If $Y(t)$ is any solution of (6.25), then $Y(t)$ is a linear combination of $Y_1, Y_2, \ldots, Y_n$. That is there exist constants $c_1, c_2, \ldots, c_n$ such that

$$Y(t) = c_1 Y_1(t) + c_2 Y_2(t) + \ldots + c_n Y_n(t), \tag{6.29}$$

for all t in J.

Proof

Let t_0 be a fixed point in J. Let $Y(t)$ be a solution of (6.25) that satisfies $Y(t_0) = Y_0$. Now, we want to show that there exist constants $c_1, c_2, \ldots, c_n$ such that the solution

$$\tilde{Y}(t) = c_1 Y_1(t) + c_2 Y_2(t) + \ldots + c_n Y_n(t),$$

satisfies

$$\tilde{Y}(t_0) = Y_0. \tag{6.30}$$

If so, then the uniqueness of the solution (Theorem 6.2.8) gives $Y(t) = \tilde{Y}(t)$ which proves (6.29). Indeed, we have

$$\tilde{Y}(t) = M(t)C,$$

where C is the vector

$$C = \begin{bmatrix} c_1 \\ c_2 \\ \vdots \\ c_n \end{bmatrix}$$

and $M(t)$ is the matrix

$$M(t) = \left[Y_1(t), Y_2(t), \ldots, Y_n(t) \right]$$

which is invertible since $Y_1(t), Y_2(t), \ldots, Y_n(t)$ are linearly independent. Thus, we have

$$\tilde{Y}(t_0) = Y_0 = M(t_0)C. \tag{6.31}$$

Hence, the vector C is given by

$$C = M^{-1}(t_0)Y_0,$$

satisfies (6.31) as desired.

Next, it is clear that the second order equation (3.7) can be written as a first order system, using the change of variables

$$y_1(t) = y(t) \qquad \text{and} \qquad y_2(t) = y'(t),$$

to get

$$\begin{cases} y_1'(t) = y_2, \\ y_2'(t) = -q(t)y_1(t) - p(t)y_2(t). \end{cases}$$

Or in the matrix form

$$\frac{d}{dt}Y(t) = A(t)Y(t),$$

with

$$Y(t) = \begin{bmatrix} y_1(t) \\ y_2(t) \end{bmatrix} \quad \text{and} \quad A(t) = \begin{bmatrix} 0 & 1 \\ -q(t) & -p(t) \end{bmatrix}.$$

We see that $\operatorname{tr} A = -p(t)$. Thus, the Wronskian in (3.8) can be rewritten as

$$W(t) = C \exp\left\{\int \operatorname{tr} A(t)dt\right\}.$$

In fact this remain also true if $A(t)$ is an $n \times n$ matrix. Thus, we extend Abel's formula (Theorem 3.1.3) as follows:

Theorem 6.2.4 (Liouville's formula)
Let us consider the linear system

$$\frac{d}{dt} Y(t) = A(t)Y(t), \tag{6.32}$$

where $Y(t)$ and $A(t)$ are continuous on an interval J and defined as in (6.10) and (6.11). Let $Y_1(t), Y_2(t), \ldots, Y_n(t)$ be n solutions of (6.32), and let t_0 in J. Then, the Wronskian of $Y_1(t), Y_2(t), \ldots, Y_n(t)$ is:

$$W[Y_1, Y_2, \ldots, Y_n](t) = W(t) = W(t_0) \exp\left\{\int_{t_0}^{t} \operatorname{tr} A(s)ds\right\}. \tag{6.33}$$

To prove (6.33), we need to show that $W(t)$ satisfies the differential equation

$$\frac{dW(t)}{dt} = \operatorname{tr} A(t)W(t),$$

form which the conclusion follows. We leave the details to the reader.

6.2.1 Fundamental and Resolvent Matrices

As we have seen above, the solution $Y(t)$ in (6.29) can be rewritten in the matrix from as

$$Y(t) = M(t)C,$$

where C is the vector

$$C = \begin{bmatrix} c_1 \\ c_2 \\ \vdots \\ c_n \end{bmatrix}. \tag{6.34}$$

Definition 6.2.3 (Fundamental matrix)

The set of n independent solutions $Y_1(t), Y_2(t), \ldots, Y_n(t)$ of the system (6.25) is called *the fundamental set* of solutions and the matrix

$$M(t) = \left[Y_1(t), Y_2(t), \ldots, Y_n(t) \right]$$

is called the *fundamental matrix* associated to (6.25).

ℹ️ **Proposition 6.2.5 (Properties of the fundamental matrix)** The fundamental matrix $M(t)$ satisfies the differential equation

$$M'(t) = A(t)M(t). \tag{6.35}$$

In addition, the fundamental matrix is not unique and if $\tilde{M}(t)$ is any other fundamental matrix of system (6.25), then we have for any t in J

$$\tilde{M}(t) = M(t)U, \tag{6.36}$$

where U is an $n \times n$ constant invertible matrix.

Proof

Formula (6.35) is a consequence of the definition of $M(t)$ and comes from the fact that each columns vector Y_i, $1 \leq i \leq n$ is a solution to the system (6.25).

Since $M(t)$ is the fundamental matrix, and if we denote by $X_1(t), X_2(t), \ldots, X_n(t)$ the vectors columns of $\tilde{M}(t)$ and since each $X_i(t)$, $1 \leq i \leq n$ is a solution of (6.25), then Theorem 6.2.3 leads to

$$X_i(t) = u_{i1} Y_1(t) + u_{i2} Y_2(t) + \ldots + u_{in} Y_n(t), \quad 1 \leq i \leq n,$$

where u_{ij}, $1 \leq i, j \leq n$ are constants . Thus, it is enough to take the matrix U whose entries are u_{ij}, $1 \leq i, j \leq n$. This proves (6.36).

Observe that for all t_0 in J, then $M(t_0)$ is an invertible matrix and $M^{-1}(t_0)$ exists since the columns of $M(t_0)$ are linearly independent. Thus, we have the following definition

---

Definition 6.2.4 (Resolvent Matrix)

Let t_0 in J. Then, the *resolvent* (or *principle fundamental matrix*) of (6.25) is defined by

$$R(t, t_0) = M(t)M^{-1}(t_0). \tag{6.37}$$

Example 6.6

Find the resolvent matrix associated to system (6.14).

Solution

The fundamental matrix associated to system (6.14) is

$$M(t) = \begin{bmatrix} e^{3t} & e^{2t} \\ e^{3t} & 2e^{2t} \end{bmatrix}.$$

To find the resolving matrix, we need to find $M^{-1}(t_0)$ for any t_0 in $\mathbb{R}$. Indeed, if

$$M^{-1}(t_0) = \begin{bmatrix} a & b \\ c & d \end{bmatrix},$$

then, using the identity

$$M(t_0)M^{-1}(t_0) = I, \tag{6.38}$$

where I is the matrix

$$I = \begin{bmatrix} 1 & 0 \\ 0 & 1 \end{bmatrix},$$

we get the system

$$\begin{cases} ae^{3t_0} + ce^{2t_0} = 1, \\ ae^{3t_0} + 2ce^{2t_0} = 0, \\ be^{3t_0} + de^{2t_0} = 0, \\ be^{3t} + 2de^{2t_0} = 1. \end{cases}$$

Solving the above system, we find

$$a = 2e^{-3t_0}, \quad b = -e^{-3t_0}, \quad c = -e^{-2t_0}, \quad d = e^{-2t_0}.$$

Thus, we obtain

$$M^{-1}(t_0) = \begin{bmatrix} 2e^{-3t_0} & -e^{-3t_0} \\ -e^{-2t_0} & e^{-2t_0} \end{bmatrix}.$$

Consequently, the resolvent matrix is

$$R(t, t_0) = M(t)M^{-1}(t_0)$$
$$= \begin{bmatrix} 2e^{-3(t-t_0)} - e^{2(t-t_0)} & -e^{3(t-t_0)} + e^{2(t-t_0)} \\ 2e^{3(t-t_0)} - 2e^{2(t-t_0)} & -e^{3(t-t_0)} + 2e^{2(t-t_0)} \end{bmatrix}.$$

Now, let s be in J, then we may also define $R(t, s)$ as

$$R(t, s) = M(t)M^{-1}(s) \tag{6.39}$$

and it satisfies the following properties.

① Proposition 6.2.6 (Properties of the resolvent matrix) The resolvent matrix satisfies for all s, t, r in J:

$$R(s, s) = I, \tag{6.40a}$$
$$R(t, s)R(s, r) = R(t, r), \tag{6.40b}$$
$$R(t, s)R(s, t) = I, \tag{6.40c}$$

In addition the matrix mapping $(t, s) \mapsto R(t, s)$ is continuous in $J \times J$ and $\partial R(t, s)/\partial t$, $\partial R(t, s)/\partial s$ exist and continuous in $J \times J$ with

$$\frac{\partial}{\partial t} R(t, s) = A(t) R(t, s) \tag{6.41a}$$

$$\frac{\partial}{\partial s} R(t, s) = -R(t, s) A(s). \tag{6.41b}$$

Proof

Formula (6.40a) can be easily seen from the definition of $R(t, s)$ as

$$R(s, s) = M(s) M^{-1}(s) = I.$$

Also, (6.40b) can be shown as follows:

$$
\begin{aligned}
R(t, s) R(s, r) &= \left(M(t) M^{-1}(s) \right) \left(M(s) M^{-1}(r) \right) \\
&= M(t) \left(M^{-1}(s) M(s) \right) M^{-1}(r) \\
&= M(t) \cdot I \cdot M^{-1}(r) \\
&= M(t) M^{-1}(r) = R(t, r).
\end{aligned}
$$

Similarly, we may simply prove (6.40c) as

$$
\begin{aligned}
R(t, s) R(s, t) &= \left(M(t) M^{-1}(s) \right) \left(M(s) M^{-1}(t) \right) \\
&= M(t) \left(M^{-1}(s) M(s) \right) M^{-1}(t) \\
&= M(t) M^{-1}(t) \\
&= I.
\end{aligned}
$$

It is also clear from (6.40c) that

$$R(s, t) = R^{-1}(t, s). \tag{6.42}$$

Now, to show (6.41), we can see first that $t \mapsto M(t)$ is differentiable, thus the mapping $s \mapsto M^{-1}(s)$ is continuous and differentiable. This, gives the continuity of $(t, s) \mapsto R(t, s)$. The existence of the partial derivatives and their continuity can be easily seen form (6.41a) and (6.41b). Thus, we need just to prove the identities (6.41a) and (6.41b). Indeed, using (6.35), we get

$$
\begin{aligned}
\frac{\partial}{\partial t} R(t, s) &= \frac{\partial}{\partial t} (M(t) M^{-1}(s)) \\
&= M'(t) M^{-1}(s) \\
&= A(t) M(t) M^{-1}(s) \\
&= A(t) R(t, s),
\end{aligned}
$$

which proves (6.41a).

On the other hand, it is not hard to show that

$$M^{-1'}(t) = -M^{-1}(t) M'(t) M^{-1}(t).$$

Using the above formula together with (6.42), we have

$$
\begin{aligned}
\frac{\partial}{\partial s} R(t,s) &= \frac{\partial}{\partial s} R^{-1}(s,t) \\
&= -R^{-1}(s,t) \cdot \frac{\partial}{\partial s} R(s,t) \cdot R^{-1}(s,t) \\
&= -R^{-1}(s,t) \cdot A(s) R(s,t) \cdot R^{-1}(s,t) \\
&= -R(t,s) A(s),
\end{aligned}
$$

which is exactly (6.41b).

Now, we have the following theorem.

Theorem 6.2.7

We consider the initial value problem

$$
\begin{cases}
Y'(t) = A(t) Y(t), \\
Y(t_0) = Y_0,
\end{cases}
\tag{6.43}
$$

where $A(t)$ is continuous on a certain open interval J that contains t_0. Then the solution of (6.43) is given by

$$
Y(t) = R(t, t_0) Y_0.
\tag{6.44}
$$

Proof

It is enough to show $Y(t) = R(t, t_0) Y_0$ satisfies (6.43), then the uniqueness theorem (Theorem 6.2.8) gives the desired result. It is clear form (6.40a), that

$$
\begin{aligned}
Y(t_0) &= R(t_0, t_0) Y_0 \\
&= I Y_0 = Y_0.
\end{aligned}
$$

On the other hand, we have from (6.41a)

$$
\begin{aligned}
\frac{d}{dt} Y(t) &= \frac{\partial}{\partial t} R(t, t_0) Y_0 \\
&= A(t) R(t, t_0) Y_0 \\
&= A(t) Y(t).
\end{aligned}
$$

Consequently the solution $Y(t) = R(t, t_0)$ satisfies (6.44). This proves Theorem 6.2.7.

Non-Homogeneous Linear System Now, we consider the non-homogenous initial value problem

$$
\begin{cases}
Y'(t) = A(t) Y(t) + B(t), \\
Y(t_0) = Y_0,
\end{cases}
\tag{6.45}
$$

where $A(t)$ and $B(t)$ are continuous on a certain open interval J that contains t_0.

> **Theorem 6.2.8 (Existence and uniqueness)**
> Assume that $A(t)$ and $B(t)$ are continuous on a certain open interval J that contains t_0. Then for any choice of the initial value Y_0, there exists a unique solution $Y(t)$ on the whole interval J to the initial value problem (6.45).

We will not prove the existence in the above theorem since its prove can be even done in the general case of system (6.1) with given initial data and under appropriate assumptions of the functions f_i, $1 \leq i \leq n$. We show only the uniqueness which is essentially based on the next lemma which is known as the *Gronwall lemma*.

6.2.2 The Gronwall Lemma

The Gronwall lemma is an important tool to prove the uniqueness of the solutions and to obtain various estimates in the theory of ordinary, partial differential equations and integral equations. There are two forms of the lemma: an integral form and a differential form. Since we will use only the integral form, we introduce just this form here.

> **Lemma 6.2.9 (Gronwall's lemma (integral form))** Let $\varphi(t)$ and $\psi(t)$ be two positive continuous functions defined on an interval J of $\mathbb{R}$. Let t_0 be in J and let $c \geq 0$ be a constant. Assume that for all t in J, we have
>
> $$\psi(t) \leq c + \left| \int_{t_0}^{t} \varphi(s)\psi(s)ds \right|. \tag{6.46}$$
>
> Then, for all t in J, we have
>
> $$\psi(t) \leq c \exp\left\{ \left| \int_{t_0}^{t} \varphi(s)ds \right| \right\}. \tag{6.47}$$
>
> In particular, if $c = 0$, then $\psi(t) = 0$, for all t in J.

Proof
Let us first assume that $c > 0$. We will discuss the two cases, $t \geq t_0$ and $t < t_0$.
For $t \geq t_0$, then we put

$$F(t) = c + \int_{t_0}^{t} \varphi(s)\psi(s)ds. \tag{6.48}$$

Hence, we have from (6.46)

$$\psi(t) \leq F(t) \tag{6.49}$$

for all t in J. On the other hand, taking the derivative of (6.48) with respect to t, we obtain

$$F'(t) = \varphi(t)\psi(t)$$
$$\leq \varphi(t)F(t), \tag{6.50}$$

where we have used (6.49). Since $F(t)$ and $F'(t)$ are positive and $F(t_0) = c$, then, we get from (6.50)

$$\ln F(t) - \ln c = \int_{t_0}^{t} \frac{F'(s)}{F(s)} ds \tag{6.51}$$

$$\leq \int_{t_0}^{t} \varphi(s)ds, \tag{6.52}$$

for all t in J. Consequently, we obtain from (6.49) and (6.51)

$$\psi(t) \leq F(t) \leq c \exp\left\{\int_{t_0}^{t} \varphi(t)\right\}$$

which is (6.47) for $t \geq t_0$.

Next, for $t < t_0$, we put

$$F(t) = c + \int_{t}^{t_0} \varphi(s)\psi(s)ds. \tag{6.53}$$

Then, we get easily,

$$F'(t) = -\varphi(t)\psi(t) \geq -\varphi(t)F(t).$$

Thus, we deduce

$$\frac{F'(t)}{F(t)} \geq -\varphi(t).$$

Integrating the above inequality, we get

$$\ln c - \ln F(t) = \int_{t}^{t_0} \frac{F'(s)}{F(s)} ds$$

$$\geq -\int_{t}^{t_0} \varphi(s)ds.$$

Consequently, we obtain

$$\psi(t) \leq F(t) \leq c \exp\left\{\int_{t}^{t_0} \varphi(t)\right\}$$

$$= c \exp\left\{\left|\int_{t_0}^{t} \varphi(s)ds\right|\right\}.$$

For $c = 0$, it is obvious that $\psi \equiv 0$.

Proof of Theorem 6.2.8 – the uniqueness

Let $Y_1(t)$ and $Y_2(t)$ be two solutions of (6.45) on the interval J. Then, the function $Y(t) = Y_1(t) - Y_2(t)$ satisfies the equation

$$\begin{cases} Y'(t) = A(t)Y(t), \\ Y(t_0) = 0. \end{cases} \tag{6.54}$$

It is clear from (6.54) that

$$Y(t) = Y(t_0) + \int_{t_0}^{t} A(s)Y(s)ds.$$

This gives

$$|Y(t)| \leq |Y(t_0)| + \int_{t_0}^{t} |A(s)||Y(s)|ds.$$

Applying Lemma 6.2.9, with $\varphi(t) = |A(t)|$, $\psi(t) = |Y(t)|$ and $c = |Y(t_0)| = 0$, we get $|Y(t)| = 0$, for all t in J. Thus, $Y \equiv 0$. This leads to the uniqueness of the solution.

Next, we prove a formula for the solution of the non-homogeneous system (6.9) with initial condition.

Theorem 6.2.10 (Variation of constants formula)
Assume that $A(t)$ and $B(t)$ are continuous on a certain open interval J that contains t_0. Then the unique solution of (6.45) is given by the formula

$$Y(t) = R(t, t_0)Y_0 + \int_{t_0}^{t} R(t, s)B(s)ds, \tag{6.55}$$

where $R(t, s)$ is the resolvent matrix associated to (6.25).

Proof
In order to show (6.55), we use the method of variation of constants similar to the one used in ▶ Sect. 3.3.3. So, since the solution of the homogeneous equation (6.25) is written as

$$Y(t) = M(t)C,$$

where $M(t)$ is the fundamental matrix associated to (6.25) and C is the vector defined in (6.34). Now, the idea is to look to C as a function of t. Thus, we need to find $Y(t)$ as a solution of (6.45) written in the form

$$Y(t) = M(t)C(t). \tag{6.56}$$

Taking the derivative of (6.56) with respect to t, we obtain

$$Y'(t) = M'(t)C(t) + M(t)C'(t).$$

Exploiting (6.35), we get

$$
\begin{aligned}
Y'(t) &= A(t)M(t)C(t) + M(t)C'(t) \\
&= A(t)Y(t) + B(t) \\
&= A(t)M(t)C(t) + B(t).
\end{aligned}
$$

Consequently, we obtain form the above formula that

$$C'(t) = M^{-1}(t)B(t). \tag{6.57}$$

Integrating (6.57) from t_0 to t, we obtain

$$C(t) = C(t_0) + \int_{t_0}^{t} M^{-1}(s)B(s)ds.$$

Using (6.56) once again, we get

$$Y_0 = Y(t_0) = M(t_0)C(t_0).$$

Thus,

$$C(t_0) = M^{-1}(t_0)Y_0.$$

Consequently, we get

$$C(t) = M^{-1}(t_0)Y_0 + \int_{t_0}^{t} M^{-1}(s)B(s)ds. \tag{6.58}$$

Inserting (6.58) into (6.56), we deduce that the solution of (6.45) is

$$
\begin{aligned}
Y(t) &= M(t)M^{-1}(t_0)Y_0 + \int_{t_0}^{t} M(t)M^{-1}(s)B(s)ds \\
&= R(t,t_0)Y_0 + \int_{t_0}^{t} R(t,s)B(s)ds,
\end{aligned}
$$

where we have used (6.37). This ends the proof of Theorem 6.2.10.

6.2.3 Linear System with Constant Coefficients

In this subsection, we investigate the homogeneous system

$$\frac{d}{dt}Y(t) = AY(t), \tag{6.59}$$

where A is an $n \times n$ constant matrix. As we have already seen, the eigenvalues and eigenvectors of the matrix A play a central role in the process of solving linear systems of differential equations. So, now, we recall (without proof) some important results related to the eigenvalues and eigenvectors of an $n \times n$ matrix.

Definition 6.2.5 (Eigenvalue and eigenvector)

A vector V is an *eigenvector* of an $n \times n$ matrix A if V is a nonzero solution to the system of linear equations

$$(A - \lambda I)V = 0. \tag{6.60}$$

The quantity λ is called an *eigenvalue* of A and V is an eigenvector associated to λ.

It is clear that form (6.60) and since V is a nonzero vector, then λ is an eigenvalue of A if and only if λ is the root of the *characteristic* equation

$$\det(A - \lambda I) = 0. \tag{6.61}$$

Since A is an $n \times n$ matrix, then (6.61) is a polynomial equation of degree n, which therefore has exactly n roots (counted with multiplicity).

Proposition 6.2.11 Suppose that $\lambda_1, \ldots, \lambda_\ell$ are real and distinct[1] eigenvalues of A with associated eigenvectors $V_1, V_2, \ldots, V_\ell$. Then, $V_1, V_2, \ldots, V_\ell$ are linearly independent.

The following corollary will play an important role in solving differential equations with constant coefficients.

Corollary 6.2.12 Suppose that A is an $n \times n$ matrix with n real distinct eigenvalues. Then, there is an invertible matrix T such that

$$T^{-1}AT = \begin{bmatrix} \lambda_1 & 0 & \cdots & 0 \\ 0 & \lambda_2 & \cdots & 0 \\ \vdots & \vdots & & \vdots \\ 0 & 0 & \cdots & \lambda_n \end{bmatrix}. \tag{6.62}$$

Since the eigenvalues are distinct, then the matrix T in Corollary 6.2.12 is defined by

$$T = \begin{bmatrix} V_1, V_2, \ldots, V_n \end{bmatrix}$$

where V_i, $1 \leq i \leq n$ is the eigenvector associated to the eigenvalue λ_i.

Example 6.7
Find the matrix T if

$$A = \begin{bmatrix} 1 & 2 & -1 \\ 0 & 3 & -2 \\ 0 & 2 & -2 \end{bmatrix}.$$

---

[1] Here "distinct" means that no two of the eigenvalues are equal.

Solution

First, we need to compute the eigenvalues of the matrix A. Thus, the characteristic equation is

$$\det(A - \lambda I) = 0 = \det \begin{bmatrix} 1-\lambda & 2 & -1 \\ 0 & 3-\lambda & -2 \\ 0 & 2 & -2-\lambda \end{bmatrix}$$

$$= (\lambda - 2)(1 - \lambda)(\lambda + 1).$$

Hence, the eigenvalues of A are

$$\lambda_1 = 2, \qquad \lambda_2 = 1, \qquad \lambda_3 = -1.$$

Now, we need to find the eigenvectors associate to the above eigenvalues. Let V_1 be the vector

$$V_1 = \begin{bmatrix} x \\ y \\ z \end{bmatrix}.$$

If, V_1 is the eigenvector associated to $\lambda_1 = 2$, then, we have

$$(A - 2I)V_1 = 0.$$

The above equation can be rewritten as

$$\begin{cases} -x + 2y - z = 0, \\ y - 2z = 0, \\ 2y - 4z = 0. \end{cases}$$

This gives $x = 3z$, $y = 2z$. Then, we may choose V_1 as

$$V_1 = \begin{bmatrix} 3 \\ 2 \\ 1 \end{bmatrix}.$$

Similarly, we may find

$$V_2 = \begin{bmatrix} 1 \\ 0 \\ 0 \end{bmatrix}, \qquad V_3 = \begin{bmatrix} 0 \\ 1 \\ 2 \end{bmatrix}$$

are the eigenvectors associated to $\lambda_2 = 1$ and $\lambda_3 = -1$, respectively. Consequently, the matrix T has the form

$$T = \begin{bmatrix} 3 & 1 & 0 \\ 2 & 0 & 1 \\ 1 & 0 & 2 \end{bmatrix}.$$

It is not hard to see that if λ is an eigenvalue of the matrix A and V is its associated eigenvector, then

$$Y(t) = Ve^{\lambda t}$$

is a nontrivial solution of (6.59).

6.2.4 Exponential of a Matrix

It is clear that equation (6.59) can be seen as a separable equation written in the matrix form. Thus, its solution is

$$Y(t) = e^{tA}C$$

where C is a vector as in (6.34). Therefore a fundamental matrix associated to (6.59) is

$$M(t) = e^{tA}, \tag{6.63}$$

which is the unique solution of the matrix equation (initial value problem)

$$M'(t) = A(t)M(t), \qquad M(0) = I.$$

Consequently, the solution of the system (6.59) is reduced to the computation of e^{tA}. Therefore, all we need to know in order to solve (6.59) is how to exponentiate a matrix.

As we have seen in ▶ Chapter 5, if a is a constant, then the function e^{at} can be represented as a power series

$$e^{at} = \sum_{k=0}^{\infty} \frac{a^k t^k}{k!}$$

which converges for all t. Now, we extend the above definition to the case where A is an $n \times n$ constant matrix.

Definition 6.2.6 (Matrix exponential)

Let A be an $n \times n$ matrix. We define the exponential of A to be the matrix given by

$$e^{tA} = \sum_{k=0}^{\infty} \frac{t^k A^k}{k!} = I + tA + \frac{t^2 A^2}{2!} + \ldots + \frac{t^k A^k}{k!} + \ldots \tag{6.64}$$

where I is the $n \times n$ identity matrix.

We may prove, by using some matrix estimates, that the series in (6.64) converges absolutely for all t in $\mathbb{R}$ and the matrix $M(t)$ in (6.63) does satisfy (6.35).

Example 6.8
Find e^{tA} for

$$A = \begin{bmatrix} 0 & 1 \\ 0 & 0 \end{bmatrix}.$$

Solution
First, we have

$$A^2 = \begin{bmatrix} 0 & 0 \\ 0 & 0 \end{bmatrix}.$$

Hence, for all $k \geq 2$, we have

$$A^k = \begin{bmatrix} 0 & 0 \\ 0 & 0 \end{bmatrix}, \qquad k \geq 2.$$

Therefore,

$$e^{tA} = I + tA = \begin{bmatrix} 1 & t \\ 0 & 1 \end{bmatrix}.$$

Example 6.9

Find e^{tA} for

$$A = \begin{bmatrix} \alpha & 0 \\ 0 & \beta \end{bmatrix},$$

where α and β are constants.

Solution

It is easy to see that

$$A^k = \begin{bmatrix} \alpha^k & 0 \\ 0 & \beta^k \end{bmatrix}.$$

Therefore,

$$e^{tA} = \begin{bmatrix} \displaystyle\sum_{k=0}^{\infty} \frac{t^k \alpha^k}{k!} & 0 \\ 0 & \displaystyle\sum_{k=0}^{\infty} \frac{t^k \beta^k}{k!} \end{bmatrix} = \begin{bmatrix} e^{\alpha t} & 0 \\ 0 & e^{\beta t} \end{bmatrix}.$$

Example 6.10

Find the fundamental matrix associated to the system of differential equations

$$\begin{cases} x'(t) = y(t), \\ y'(t) = x(t). \end{cases} \tag{6.65}$$

Solution

System (6.65) can be easily written in the matrix form

$$\frac{d}{dt} Y(t) = AY(t),$$

with

$$Y(t) = \begin{bmatrix} x(t) \\ y(t) \end{bmatrix} \qquad \text{and} \qquad A = \begin{bmatrix} 0 & 1 \\ 1 & 0 \end{bmatrix}.$$

So, according to what we have seen above, to find a fundamental matrix $M(t)$ associated to (6.65), then we need to compute e^{tA}. Indeed, we have

$$A^2 = \begin{bmatrix} 1 & 0 \\ 0 & 1 \end{bmatrix} \quad \text{and} \quad A^3 = A \cdot A^2 = A \cdot I = A.$$

Thus, we may easily show that for $k \geq 0$, we have

$$A^{2k} = I \quad \text{and} \quad A^{2k+1} = A.$$

Consequently, the fundamental matrix $M(t)$ associated to (6.65) is

$$M(t) = e^{tA} = I + tA + \frac{t^2 A^2}{2!} + \ldots + \frac{t^k A^k}{k!} + \ldots$$

$$= \begin{bmatrix} 1 + \dfrac{t^2}{2!} + \ldots + \dfrac{t^{2k}}{(2k)!} + \ldots & t + \dfrac{t^3}{3!} + \ldots + \dfrac{t^{2k+1}}{(2k+2)!} + \ldots \\ t + \dfrac{t^3}{3!} + \ldots + \dfrac{t^{2k+1}}{(2k+2)!} + \ldots & 1 + \dfrac{t^2}{2!} + \ldots + \dfrac{t^{2k}}{(2k)!} + \ldots \end{bmatrix}$$

$$= \begin{bmatrix} \cosh t & \sinh t \\ \sinh t & \cosh t \end{bmatrix}.$$

The following proposition shows that e^{tA} shares many of the familiar properties of the usual exponential function.

ⓘ Proposition 6.2.13 (Properties of the matrix exponential) Let A, B and T be $n \times n$ matrices. Then, we have:

1. If $B = T^{-1}AT$, then

    $$e^{tB} = T^{-1} e^{tA} T.$$

2. If $AB = BA$, then

    $$e^{t(A+B)} = e^{tA} e^{tB}.$$

3. The matrix e^{tA} is invertible and

    $$(e^{tA})^{-1} = e^{-tA}.$$

4. The matrix e^{tA} is differentiable for all t in $\mathbb{R}$ and

    $$\frac{d}{dt} e^{tA} = A e^{tA} = e^{tA} A.$$

The proof of Proposition 6.2.13 can be done using the properties of matrices and power series. In addition by knowing the form of the resolvent matrix (6.66), we can easily see that the above properties are direct application of those in Proposition 6.2.6.

ⓘ Proposition 6.2.14 The resolvent matrix associated to system (6.59) is

$$R(t, s) = e^{(t-s)A}. \tag{6.66}$$

Proof

The proof of (6.66) can be seen easily from (6.37), (6.63) and Proposition 6.2.13 as

$$R(t, s) = M(t)M^{-1}(s)$$
$$= e^{tA}e^{-sA} = e^{(t-s)A}.$$

6.2.5 Computation of the Matrix Exponential

The exponential of a matrix can be computed in many ways. As we have seen in ▶ Sect. 6.2.4, we can use the definition of e^{tA} to compute it, where neither the eigenvalues nor the eigenvectors of the matrix play a direct role in the actual computations. But this method is very hard to apply in general even for 2×2 matrices.

Here we present the classical method based on the matrix eigenvalues and eigenvectors. We also introduced a method, which is not very well known and requires information only on the eigenvalues and not on the eigenvectors of the matrix A.

As we have seen in Example 6.9, if A is diagonal matrix, that is

$$A = \mathrm{diag}(\lambda_1, \lambda_2, \ldots, \lambda_n) = \begin{bmatrix} \lambda_1 & 0 & \cdots & 0 \\ 0 & \lambda_2 & \cdots & 0 \\ \vdots & \vdots & & \vdots \\ 0 & 0 & \cdots & \lambda_n \end{bmatrix},$$

then we get for any $k \geq 0$,

$$A^k = \mathrm{diag}(\lambda_1^k, \lambda_2^k, \ldots, \lambda_n^k) = \begin{bmatrix} \lambda_1^k & 0 & \cdots & 0 \\ 0 & \lambda_2^k & \cdots & 0 \\ \vdots & \vdots & & \vdots \\ 0 & 0 & \cdots & \lambda_n^k \end{bmatrix}.$$

Then, using Definition 6.64, we obtain

$$e^{tA} = \mathrm{diag}(e^{t\lambda_1}, e^{t\lambda_2}, \ldots, e^{t\lambda_n}) = \begin{bmatrix} e^{t\lambda_1} & 0 & \cdots & 0 \\ 0 & e^{t\lambda_2} & \cdots & 0 \\ \vdots & \vdots & & \vdots \\ 0 & 0 & \cdots & e^{t\lambda_n} \end{bmatrix}. \tag{6.67}$$

This is the easiest case, where we can find e^{tA}.

Another case is when we can write the matrix A as the sum of two matrices

$$A = D + N,$$

where D is a diagonal matrix such that all its entries are equal:

$$D = \begin{bmatrix} \lambda & 0 & \cdots & 0 \\ 0 & \lambda & \cdots & 0 \\ \vdots & \vdots & & \vdots \\ 0 & 0 & \cdots & \lambda \end{bmatrix} = \lambda I,$$

where I is the $n \times n$ identity matrix, and N (called the *nilpotent* matrix) satisfies

$$N^{k_0} = 0,$$

(zero here is the $n \times n$ matrix with all its entries are zero) for some positive integer $k_0 > 0$. Here the simplicity of the method depends of the smallness of k_0. In this case, D commutes with any other square matrix of the same order. In particular we have $DN = ND$. Thus, applying the second formula in Proposition 6.2.13, we obtain

$$e^{tA} = e^{tN} e^{tD}$$

$$= \left(I + tN + \frac{t^2 N^2}{2!} + \ldots + \frac{t^{k_0-1} N^{k_0-1}}{(k_0 - 1)!} \right) \text{diag}(e^{t\lambda}, e^{t\lambda}, \ldots, e^{t\lambda})$$

$$= e^{t\lambda} \left(I + tN + \frac{t^2 N^2}{2!} + \ldots + \frac{t^{k_0-1} N^{k_0-1}}{(k_0 - 1)!} \right). \tag{6.68}$$

Example 6.11
Find e^{tA} for

$$A = \begin{bmatrix} a & b \\ 0 & a \end{bmatrix},$$

where a and b are constants.

Solution
The matrix A can be written as $A = D + N$ with

$$D = \begin{bmatrix} a & 0 \\ 0 & a \end{bmatrix} \quad \text{and} \quad N = \begin{bmatrix} 0 & b \\ 0 & 0 \end{bmatrix}.$$

It is clear that $N^2 = 0$. Thus, applying formula (6.68), we get

$$e^{tA} = e^{at} (I + tN)$$

$$= e^{at} \begin{bmatrix} 1 & bt \\ 0 & 1 \end{bmatrix} = \begin{bmatrix} e^{at} & bt e^{at} \\ 0 & e^{at} \end{bmatrix}.$$

Now, looking to the first property in Proposition 6.2.13, and keeping (6.67) in mind, then it is easy to see that if we can write the matrix A in the form

$$A = SDS^{-1}, \tag{6.69}$$

where D is a diagonal matrix and S is an invertible matrix, then we have

$$e^{tA} = Se^{tD}S^{-1}. \tag{6.70}$$

Any matrix A which satisfies (6.70) is called *diagonalizable* matrix.

Our main goal now is to look for the matrices which satisfy (6.70). To diagonalize A, we need n linearly independent eigenvectors of A. Otherwise this method of computing the exponential does not work.

Corollary 6.2.12 shows that (6.70) holds if the eigenvalues of A are simple and in this case $S = T^{-1}$, where T is the matrix where its columns are the eigenvectors of A. In this case the eigenvalues of A are the entries of the diagonal matrix D. Therefore any matrix with distinct eigenvalues can be diagonalizable. It is clear that the matrix S is not unique, since an eigenvector V can be multiplied by a constant and will remain an eigenvector. Before discussing the other cases, let us give the following definitions.

Definition 6.2.7 (Algebraic and geometric multiplicities)

The *algebraic multiplicity* of an eigenvalue λ is the number of times λ appears as a root of the characteristic equation

$$\det(A - \lambda I) = 0.$$

The *geometric multiplicity* of an eigenvalue λ is the number of independent eigenvectors associated to λ.

Definition 6.2.8 (Complete and defective eigenvalues)

Assume that A is an $n \times n$ matrix and λ is a repeated eigenvalue of A with multiplicity m. Then:

- the eigenvalue λ is called *complete* if there are m linearly independent eigenvectors corresponding to it. That is if the geometric multiplicity is equal to the algebraic multiplicity. In this case the matrix A is diagonalizable.
- The eigenvalue λ is *defective* if the geometric multiplicity is strictly less than the algebraic multiplicity. In this case the matrix A is called a *defective matrix* and it is not diagonalizable.

Example 6.12

Show that the matrix

$$A = \begin{bmatrix} -2 & 1 & 1 \\ 1 & -2 & 1 \\ 1 & 1 & -2 \end{bmatrix},$$

is diagonalizable.

Solution

We need first to find the eigenvalues of the matrix A by solving the characteristic equation

$$\det(A - \lambda I) = 0. \tag{6.71}$$

It is not hard to see that

$$\det(A - \lambda I) = \det \begin{bmatrix} -2 - \lambda & 1 & 1 \\ 1 & -2 - \lambda & 1 \\ 1 & 1 & -2 - \lambda \end{bmatrix}$$

$$= \lambda^3 + 6\lambda^2 + 9\lambda.$$

Thus, the solutions of (6.71) are

$$\lambda_1 = 0, \qquad \lambda_2 = -3.$$

It is clear that λ_2 has an algebraic multiplicity 2 (that is $m = 2$). Now, in order for λ_2 to be complete, we need to find two independent eigenvectors associated to $\lambda_2 = -3$. That is, we need to show that the geometric multiplicity is also equal to two. Indeed, let

$$V = \begin{bmatrix} x \\ y \\ z \end{bmatrix}.$$

Then, V is an eigenvector associated to $\lambda_2 = -3$, if and only if

$$(A + 3I)V = 0.$$

Or equivalently

$$\begin{cases} x + y + z = 0, \\ x + y + z = 0, \\ x + y + z = 0. \end{cases}$$

This means that V is an eigenvector associated to $\lambda = -3$ if and only if its components satisfies

$$x + y + z = 0.$$

Or if its third component z satisfies

$$z = -x - y.$$

Therefore, we may write V as

$$V = \begin{bmatrix} x \\ y \\ -x - y \end{bmatrix} = x \begin{bmatrix} 1 \\ 0 \\ -1 \end{bmatrix} + y \begin{bmatrix} 0 \\ 1 \\ -1 \end{bmatrix}.$$

Consequently, the two vectors

$$V_1 = \begin{bmatrix} 1 \\ 0 \\ -1 \end{bmatrix} \quad \text{and} \quad V_2 = \begin{bmatrix} 0 \\ 1 \\ -1 \end{bmatrix}$$

are two independent eigenvectors associated to $\lambda_2 = -3$. Therefore, the geometric multiplicity is equal to 2. Consequently, $\lambda_2 = -3$ is a complete eigenvalue and thus, the matrix A is diagonalizable.

Example 6.13

Show that the matrix

$$A = \begin{bmatrix} 0 & 1 \\ 0 & 0 \end{bmatrix},$$

is defictive.

Solution

It is clear that

$$\det(A - \lambda I) = \lambda^2.$$

Thus, $\lambda = 0$ is a an eigenvalue of multiplicity 2. If

$$V = \begin{bmatrix} x \\ y \end{bmatrix}.$$

is an eigenvector associated to λ, then $y = 0$. Thus all the eigenvalues of A are multiple of the vector

$$V = \begin{bmatrix} 1 \\ 0 \end{bmatrix}.$$

Therefore, the geometric multiplicity is 1. Consequently, the matrix A is defective and we cannot find S satisfying (6.69).

The following theorem is very useful in practice.

Theorem 6.2.15 (Eigenvalues of a symmetric matrix)

If A is an $n \times n$ real matrix which is symmetric, that is

$$A^T = A,$$

where A^T is the matrix transpose of A, then all its eigenvalues are complete. Therefore, it is diagonalizable.

Using this theorem, we may deduce directly (without any computation) that the matrix A in Example 6.12 is diagonalizable since it is symmetric.

6.2.6 A General Method of Computing Matrix Exponential

As, we have seen above, up until now, we know how to find the exponential of an $n \times n$ matrix A only in the case where the matrix has a special form or if it is diagonalizable. Even in this case, it is not easy to find all the eigenvalues and check if they are complete and then find the eigenvectors. Even, if we succeeded to do this, it will be a challenging task to find the inverse of the matrix S in (6.69). In addition, this method works only for diagonalizable matrices. So, the question is: *how can we compute the exponential of A if it is not diagonalizable?* One classical way to answer this question is to use the *Jordan canonical form*. Thus we show that there exists an invertible matrix S such that

$$A = SJS^{-1},$$ (6.72)

where J is the block diagonal matrix

$$J = \begin{bmatrix} J_1 & 0 & \cdots & 0 \\ 0 & J_2 & \cdots & 0 \\ \vdots & \vdots & & \vdots \\ 0 & 0 & \cdots & J_\ell \end{bmatrix}$$

where J_i, $1 \le i \le \ell \le n$ are the Jordan blocks. That is for a fixed i, the J_i is a $k_i \times k_i$ matrix associated to an eigenvalue λ_i and has the form

$$J_i = \begin{bmatrix} \lambda_i & 1 & & 0 \\ 0 & \lambda_i & \ddots & \\ \vdots & \vdots & \ddots & 1 \\ 0 & 0 & \cdots & \lambda_i \end{bmatrix}.$$

In other words, a Jordan block is a matrix with all the diagonal entries equal to each other, all the entries immediately above the diagonal equal to 1 and all the other entries equal to 0. For example every diagonal matrix is a matrix in the Jordan canonical form with each Jordan block is 1×1 block.

Writing A in the form (6.72) will lead eventually to the computation of e^{tA} simply by computing e^{tJ} as

$$e^{tJ} = \begin{bmatrix} e^{tJ_1} & 0 & \cdots & 0 \\ 0 & e^{tJ_2} & \cdots & 0 \\ \vdots & \vdots & & \vdots \\ 0 & 0 & \cdots & e^{tJ_\ell} \end{bmatrix}$$

and hence it remains to compute the exponential of each Jordan block. However, the method of obtaining the matrix J requires deep knowledge of *generalized* eigenvector for each eigenvalue of A, which is a vector V satisfying

$$(A - \lambda I)^k V = 0,$$

for some positive integer k, and also requires some knowledge of the *generalized* eigenspace of λ:

$$E_\lambda = \{V \mid (A - \lambda I)^k V = 0, \text{ for some } k\}.$$

Consequently, the determination of the Jordan canonical form for a given matrix A can be difficult, if not impossible, to do in practice. So, to avoid the Jordan canonical form, we introduce an alternative method, which is not well known, and has been published in the paper [18]. This method based solely on the knowledge of the eigenvalues of A and their algebraic multiplicity, for finding e^{tA} and avoids the complicated Jordan form.

We describe the method in the following theorem.

Theorem 6.2.16

Assume that A is an $n \times n$ matrix with the eigenvalues $\lambda_1, \lambda_2, \ldots, \lambda_n$ (real or complex) written in some arbitrary but specified order and they are not necessary distinct. Then

$$e^{tA} = \sum_{j=0}^{n-1} r_{j+1}(t) P_j \tag{6.73}$$

where

$$P_0 = I, \qquad P_j = \prod_{k=1}^{j} (A - \lambda_j I), \qquad j = 1, \ldots, n$$

and $r_1(t), \ldots, r_n(t)$ are (real or complex) functions defined by the following equations

$$\begin{cases} r_1'(t) = \lambda_1 r_1(t), & r_1(0) = 1, \\ r_2'(t) = \lambda_2 r_2(t) + r_1(t), & r_2(0) = 0, \\ \vdots & \\ r_n'(t) = \lambda_n r_n(t) + r_{n-1}(t), & r_n(0) = 0. \end{cases} \tag{6.74}$$

The basic idea of this method relies on the Cayley–Hamilton theorem, where its proof can be found in many books of linear algebra.

Theorem 6.2.17 (Cayley–Hamilton theorem)

Let A be an $n \times n$ matrix and I be the $n \times n$ identity matrix. Let $p(\lambda)$ be the *characteristic polynomial*

$$p(\lambda) = \det(A - \lambda I).$$

Then,

$$p(A) = 0.$$

Proof of Theorem 6.2.16

Let us define

$$\Phi(t) = \sum_{j=0}^{n-1} r_{j+1}(t) P_j. \tag{6.75}$$

Our goal is to show that $\Phi(t)$ satisfies the initial value problem

$$\begin{cases} \Phi'(t) = A\Phi(t), \\ \Phi(0) = I. \end{cases} \tag{6.76}$$

Therefore, since e^{tA} is the unique solution of (6.76), then we deduce that

$$\Phi(t) = e^{tA}.$$

So, now we need just to show (6.76). Indeed, we have

$$\Phi(0) = r_1(0)P_0 = I.$$

On the other hand, taking the derivative of (6.75) with respect to t, we obtain

$$\Phi'(t) = \sum_{j=0}^{n-1} r'_{j+1}(t) P_j = \sum_{j=2}^{n} \{\lambda_j r_j(t) + r_{j-1}(t)\} P_{j-1} + \lambda_1 r_1(t) P_0.$$

Shifting the index and taking $r_0(t) = 0$, then we may rewrite the above formula as

$$\Phi'(t) = \sum_{j=0}^{n-1} \{\lambda_{j+1} r_{j+1}(t) + r_j(t)\} P_j.$$

Hence,

$$\Phi'(t) - \lambda_n \Phi(t) = \sum_{j=0}^{n-1} \{\lambda_{j+1} r_{j+1}(t) + r_j(t)\} P_j - \lambda_n \sum_{j=0}^{n-1} r_{j+1}(t) P_j$$

$$= \sum_{j=0}^{n-1} (\lambda_{j+1} - \lambda_n) r_{j+1}(t) P_j + \sum_{j=0}^{n-1} r_j(t) P_j$$

$$= \sum_{j=0}^{n-2} \{(\lambda_{j+1} - \lambda_n) r_{j+1}(t) P_j + r_{j+1}(t) P_{j+1}\}.$$

We put

$$P_{j+1} = (A - \lambda_{j+1} I) P_j, \qquad 0 \le j \le n-2,$$

we get

$$\Phi'(t) - \lambda_n \Phi(t) = \sum_{j=0}^{n-2} \{(\lambda_{j+1} - \lambda_n)I + A - \lambda_{j+1} I\} r_{j+1}(t) P_j$$

$$= \sum_{j=0}^{n-2} (A - \lambda_n I) r_{j+1}(t) P_j$$

$$= (A - \lambda_n)(\Phi(t) - r_n(t) P_{n-1}). \tag{6.77}$$

Applying Theorem 6.2.17, we have

$$p(A) = (A - \lambda_n) P_j = 0,$$

where $p(\lambda)$ is the characteristic polynomial defined by

$$p(\lambda) = \det(A - \lambda I) = (-1)^n (\lambda - \lambda_1) \cdots (\lambda - \lambda_n).$$

Consequently (6.76) is satisfied.

Example 6.14

Apply the method in Theorem 6.2.16 to find the solution of the following system

$$\begin{cases} x'(t) = 2x(t) + y(t), \\ y'(t) = 3x(t) + 4y(t). \end{cases} \qquad (6.78)$$

Solution

Let $Y(t)$ be the vector

$$Y(t) = \begin{bmatrix} x(t) \\ y(t) \end{bmatrix}.$$

Therefore, system (6.78) can be written as

$$\frac{d}{dt} Y(t) = AY(t),$$

with

$$A = \begin{bmatrix} 2 & 1 \\ 3 & 4 \end{bmatrix}.$$

Thus, its solution is given by

$$Y(t) = e^{tA} C,$$

where

$$C = \begin{bmatrix} c_1 \\ c_2 \end{bmatrix}.$$

Now, in order to find the matrix e^{tA}, we need first to compute the eigenvalues of A. Indeed, we have

$$\det(A - \lambda I) = (\lambda - 1)(\lambda - 5).$$

This gives the two eigenvalues $\lambda_1 = 1$ and $\lambda_2 = 5$. Now, we need to find $r_1(t)$ and $r_2(t)$ as solutions of the system

$$\begin{cases} r_1'(t) = \lambda_1 r_1(t), & r_1(0) = 1, \\ r_2'(t) = \lambda_2 r_2(t) + r_1(t), & r_2(0) = 0, \end{cases}$$

This gives

$$r_1(t) = e^{\lambda_1 t} = e^t$$

and

$$r_2(t) = \int_0^t e^{\lambda_2(t-s)} r_1(s) ds$$

$$= \int_0^t e^{5t} e^{-4s} ds = \frac{e^{5t} - e^t}{4}.$$

Now, we need to compute P_1 as

$$P_1 = A - \lambda_1 I$$

$$= A - I = \begin{bmatrix} 1 & 1 \\ 3 & 3 \end{bmatrix}.$$

Consequently applying formula (6.73), we get

$$e^{tA} = r_1(t)P_0 + r_2(t)P_1$$

$$= e^t \begin{bmatrix} 1 & 0 \\ 0 & 1 \end{bmatrix} + \frac{e^{5t} - e^t}{4} \begin{bmatrix} 1 & 1 \\ 3 & 3 \end{bmatrix}$$

$$= \begin{bmatrix} \frac{3}{4}e^t + \frac{1}{4}e^{5t} & -\frac{1}{4}e^t + \frac{1}{4}e^{5t} \\ -\frac{3}{4}e^t + \frac{3}{4}e^{5t} & \frac{1}{4}e^t + \frac{3}{4}e^{5t} \end{bmatrix}.$$

In Example 6.14 the eigenvalues were district and the matrix A is diagonalizable. Now, to show the powerful of this method, we may apply it to the case where the matrix A is defective and we show how we can obtain the solution without using the Jordan canonical form as in the following example.

Example 6.15
Find the solution of the following system

$$\begin{cases} x'(t) = -2x(t) - y(t) - 3z(t), \\ y'(t) = 4x(t) + 3y(t) + 3z(t), \\ z'(t) = -2x(t) + y(t) - z(t), \\ x(0) = 1, \quad y(0) = 0, \quad z(0) = 1. \end{cases} \tag{6.79}$$

Solution
System (6.79) can be written in the matrix form as

$$\begin{cases} Y'(t) = AY(t), \\ Y(0) = Y_0, \end{cases} \tag{6.80}$$

where

$$Y(t) = \begin{bmatrix} x(t) \\ y(t) \\ z(t) \end{bmatrix}, \quad A = \begin{bmatrix} -2 & -1 & -3 \\ 4 & 3 & 3 \\ -2 & 1 & -1 \end{bmatrix}, \quad Y_0 = \begin{bmatrix} 1 \\ 0 \\ 1 \end{bmatrix}.$$

Then, a simple computation shows that

$$\det(A - \lambda I) = (2 - \lambda)^2(-4 - \lambda).$$

Therefore the eigenvalues of A are $\lambda_1 = 2$ with algebraic multiplicity 2 and $\lambda_2 = -4$ with algebraic multiplicity 1. We may easily check that that the geometric multiplicity of λ_1 is 1 since the only linearly independent eigenvector associated to it is

$$V = \begin{bmatrix} 1 \\ -1 \\ -1 \end{bmatrix}.$$

Consequently, A is defective and therefore not diagonalizable. Recalling (6.55) and (6.66), we deduce that the solution of (6.80) is given by

$$Y(t) = e^{tA} Y_0. \tag{6.81}$$

In order to find the matrix e^{tA}, we use formula (6.73) and compute

$$r_1(t) = e^{\lambda_1 t} = e^{2t}$$

and

$$r_2(t) = \int_0^t e^{\lambda_1 (t-s)} r_1(s) ds$$

$$= \int_0^t e^{2t} ds = t e^{2t}.$$

Similarly, we get from (6.74) that

$$r_3(t) = \int_0^t e^{\lambda_2 (t-s)} r_2(s) ds$$

$$= \int_0^t e^{-4t} s e^{6s} ds$$

$$= e^{-4t} \left[\left(\frac{1}{6} s - \frac{1}{36} \right) e^{6s} \right]_0^t$$

$$= \frac{1}{36} e^{-4t} + \left(\frac{1}{6} t - \frac{1}{36} \right) e^{2t}.$$

Now, we compute P_1 and P_2 as

$$P_1 = A - 2I = \begin{bmatrix} -4 & -1 & -3 \\ 4 & 1 & 3 \\ -2 & 1 & -3 \end{bmatrix}$$

and

$$P_2 = (A - 2I)(A - 2I) = \begin{bmatrix} 18 & 0 & 18 \\ -18 & 0 & -18 \\ 18 & 0 & 18 \end{bmatrix}.$$

Consequently, formula (6.73) gives the following resolvent matrix

$$e^{tA} = r_1(t)P_0 + r_2(t)P_1 + r_3(t)P_2$$

$$= e^{2t}\begin{bmatrix} 1 & 0 & 0 \\ 0 & 1 & 0 \\ 0 & 0 & 1 \end{bmatrix} + te^{2t}\begin{bmatrix} -4 & -1 & -3 \\ 4 & 1 & 3 \\ -2 & 1 & -3 \end{bmatrix}$$

$$+ \left(\frac{1}{36}e^{-4t} + \left(\frac{1}{6}t - \frac{1}{36}\right)e^{2t}\right)\begin{bmatrix} 18 & 0 & 18 \\ -18 & 0 & -18 \\ 18 & 0 & 18 \end{bmatrix}$$

$$= \begin{bmatrix} (\frac{1}{2} - t)e^{2t} + \frac{1}{2}e^{-4t} & -te^{2t} & -\frac{1}{2}e^{2t} + \frac{1}{2}e^{-4t} \\ (t + \frac{1}{2})e^{2t} - \frac{1}{2}e^{-4t} & (1 + t)e^{2t} & \frac{1}{2}e^{2t} - \frac{1}{2}e^{-4t} \\ (t - \frac{1}{2})e^{2t} + \frac{1}{2}e^{-4t} & te^{2t} & \frac{1}{2}e^{2t} + \frac{1}{2}e^{-4t} \end{bmatrix}.$$

Consequently, keeping in mind (6.81), we get the solution of (6.80)

$$Y(t) = \begin{bmatrix} e^{-4t} - te^{2t} \\ -e^{-4t} + (t + 1)e^{2t} \\ e^{-4t} + te^{2t} \end{bmatrix}.$$

Therefore, the solution of (6.79) is

$$\begin{cases} x(t) = e^{-4t} - te^{2t}, \\ y(t) = (t + 1)e^{2t} - e^{-4t}, \\ z(t) = e^{-4t} + te^{2t}. \end{cases}$$

6.2.7 The nth-Order Linear Equation

In this subsection, we continue the discussion that we have started in ▶ Sect. 6.1 and we apply the above obtained results to the following linear differential equation of order n:

$$a_n y^{(n)}(t) + a_{n-1} y^{(n-1)}(t) + \cdots + a_1 y'(t) + a_0(t)y(t) = b(t). \tag{6.82}$$

Here $y = y(t)$ is an unknown scalar function and the coefficients $a_i(t)$, $i = 1, \ldots, n$ are continuous on some open interval J and $a_n(t) \neq 0$ for all t in J. Using the change of variables in (6.3), we can write (6.82) as a first order system of the form

$$Y'(t) = A(t)Y(t) + B(t) \tag{6.83}$$

with

$$Y(t) = \begin{bmatrix} y_1(t) \\ y_2(t) \\ \vdots \\ y_n(t) \end{bmatrix}, \qquad B(t) = \begin{bmatrix} 0 \\ 0 \\ \vdots \\ b(t)/a_n(t) \end{bmatrix} \tag{6.84}$$

and $A(t)$ is the matrix

$$A(t) = \begin{bmatrix} 0 & 1 & 0 & \cdots & 0 \\ 0 & 0 & 1 & \cdots & 0 \\ \vdots & \vdots & \vdots & \vdots & \vdots \\ 0 & 0 & \cdots & 0 & 1 \\ -a_0/a_n & -a_1/a_n & \cdots & \cdots & -a_{n-1}/a_n \end{bmatrix}. \tag{6.85}$$

Now, we claim that equation (6.82) is equivalent to system (6.83). Indeed, to a solution $y = \psi(t)$ of equation (6.82), it corresponds the vector solution

$$Y(t) = \varphi(t) = \begin{bmatrix} \psi(t) \\ \psi'(t) \\ \vdots \\ \psi^{(n-1)}(t) \end{bmatrix}$$

to system (6.83). Conversely, given a solution

$$Y(t) = \varphi(t) = \begin{bmatrix} \varphi_1(t) \\ \varphi_2(t) \\ \vdots \\ \varphi_n(t) \end{bmatrix}$$

to system (6.83), thus $y = \varphi_1(t)$ is a solution to (6.82) and system (6.83) implies

$$y_1'(t) = \varphi_1'(t) = \varphi_2(t), \ldots, y^{(n-1)}(t) = \varphi_{n-1}'(t) = \varphi_n(t).$$

Since equation (6.82) is equivalent to system (6.83), then from (6.33), we may deduce that the Wronskian of any n solutions of (6.82) is given by

$$W(t) = W(t_0) \exp\left\{ -\int_{t_0}^{t} \frac{a_{n-1}(s)}{a_n(s)} ds \right\}.$$

In the next example, we will show how to use the equivalent first order system in order to find a solution to a third order equation.

Example 6.16

Find the solution of the third order equation

$$\begin{cases} y'''(t) - 4y'(t) = t + 3\cos t + e^{-2t} \\ y(0) = 1, \quad y'(0) = 1, \quad y''(0) = 0. \end{cases} \tag{6.86}$$

Solution

According to what we have seen above, introducing the following change of variables

$$y_1 = y, \qquad y_2 = y', \qquad y_3 = y'',$$

then, equation (6.86) is equivalent to the first order linear system

$$\begin{cases} y_1'(t) = y_2(t), \\ y_2'(t) = y_3(t), \\ y_3'(t) = 4y_2(t) + t + 3\cos t + e^{-2t}, \\ y_1(0) = 1, \quad y_2(0) = 1, \quad y_3(0) = 0. \end{cases} \qquad (6.87)$$

This can be also written in the matrix form

$$\begin{cases} Y'(t) = A(t)Y(t) + B(t), \\ Y(0) = Y_0, \end{cases} \qquad (6.88)$$

with

$$Y(t) = \begin{bmatrix} y_1(t) \\ y_2(t) \\ y_3(t) \end{bmatrix}, \quad A = \begin{bmatrix} 0 & 1 & 0 \\ 0 & 0 & 1 \\ 0 & 4 & 0 \end{bmatrix}, \quad B(t) = \begin{bmatrix} 0 \\ 0 \\ t + 3\cos t + e^{-2t} \end{bmatrix},$$

and

$$Y_0 = \begin{bmatrix} 1 \\ 1 \\ 0 \end{bmatrix}.$$

Now, we need to find the resolvent matrix $R(t,0) = e^{tA}$ associated to (6.88). It is clear that

$$\det(A - \lambda I) = -\lambda(\lambda - 2)(\lambda + 2).$$

This gives that $\lambda_1 = 0$, $\lambda_2 = 2$ and $\lambda_3 = -2$ are the eigenvalues of A.

Now, to compute e^{tA}, we use formula (6.73), and compute $r_1(t)$, $r_2(t)$ and $r_3(t)$ as follows

$$r_1(t) = e^{\lambda_1 t} = 1,$$

$$r_2(t) = \int_0^t e^{\lambda_2(t-s)} r_1(s)\,ds$$

$$= \int_0^t e^{2t} e^{-2s}\,ds = \frac{e^{2t} - 1}{2}$$

and

$$r_3(t) = \int_0^t e^{\lambda_3(t-s)} r_2(s)\,ds$$

$$= \frac{1}{2}\int_0^t e^{-2t}(e^{4s} - e^{2s})\,ds$$

$$= \frac{1}{2}e^{-2t}\left[\left(\frac{1}{4}e^{4s} - \frac{1}{2}e^{2s}\right)\right]_0^t$$

$$= \frac{1}{8}e^{-2t} + \frac{1}{8}e^{2t} - \frac{1}{4}.$$

Next, we compute the matrices P_1 and P_2 as

$$P_1 = A - \lambda_1 I = A$$

and

$$P_2 = (A - \lambda_1 I)(A - \lambda_2 I) = A(A - 2I)$$

$$= \begin{bmatrix} 0 & -2 & 1 \\ 0 & 4 & -2 \\ 0 & -8 & 4 \end{bmatrix}.$$

Consequently, applying formula (6.73), we get

$$e^{tA} = r_1(t)P_0 + r_2(t)P_1 + r_3(t)P_2$$

$$= \begin{bmatrix} 1 & 0 & 0 \\ 0 & 1 & 0 \\ 0 & 0 & 1 \end{bmatrix} + \frac{e^{2t} - 1}{2} \begin{bmatrix} 0 & 1 & 0 \\ 0 & 0 & 1 \\ 0 & 4 & 0 \end{bmatrix} + \frac{1}{8}(e^{-2t} + e^{2t} - 2) \begin{bmatrix} 0 & -2 & 1 \\ 0 & 4 & -2 \\ 0 & -8 & 4 \end{bmatrix}$$

$$= \begin{bmatrix} 1 & \frac{1}{4}(e^{2t} - e^{-2t}) & \frac{1}{8}(e^{2t} + e^{-2t}) - \frac{1}{4} \\ 0 & \frac{1}{2}(e^{2t} + e^{-2t}) & \frac{1}{4}(e^{2t} - e^{-2t}) \\ 0 & e^{2t} - e^{-2t} & \frac{1}{2}(e^{2t} + e^{-2t}) \end{bmatrix}.$$

Therefore the resolvent matrix is:

$$R(t,0) = e^{tA} = \begin{bmatrix} 1 & \frac{1}{2}\sinh 2t & \frac{1}{4}\cosh 2t - \frac{1}{4} \\ 0 & \cosh 2t & \frac{1}{2}\sinh 2t \\ 0 & 2\sinh 2t & \cosh 2t \end{bmatrix}.$$

On the other hand, it is clear that

$$e^{-sA} = \begin{bmatrix} 1 & -\frac{1}{2}\sinh 2s & \frac{1}{4}\cosh 2s - \frac{1}{4} \\ 0 & \cosh 2s & -\frac{1}{2}\sinh 2s \\ 0 & -2\sinh 2s & \cosh 2s \end{bmatrix}.$$

This leads to

$$R(t,s) = e^{(t-s)A} = \begin{bmatrix} 1 & \frac{1}{2}\sinh 2(t-s) & \frac{1}{4}\cosh 2(t-s) - \frac{1}{4} \\ 0 & \cosh 2(t-s) & \frac{1}{2}\sinh 2(t-s) \\ 0 & 2\sinh 2(t-s) & \cosh 2(t-s) \end{bmatrix}.$$

Now, applying formula (6.55), we deduce that

$$Y(t) = R(t,0)Y_0 + \int_0^t R(t,s)B(s)ds. \tag{6.89}$$

Since $y_1(t) = y(t)$, then the solution of (6.86) is the first component in the vector $Y(t)$ in (6.89). That is

$$y(t) = 1 + \frac{1}{2}\sinh 2t + \int_0^t \left(\frac{1}{4}\cosh 2(t-s) - \frac{1}{4}\right)(s + 3\cos s + e^{-2s})ds$$

$$= \frac{7}{8} - \frac{t^2}{8} - \frac{3}{4}\sin t + \frac{1}{8}e^{-2t} + \frac{1}{2}\sinh 2t$$

$$+ \frac{1}{4}\int_0^t \cosh 2(t-s)(s + 3\cos s + e^{-2s})ds. \tag{6.90}$$

Using integration by parts, we get

$$\frac{1}{4}\int_0^t \cosh 2(t-s)(s + 3\cos s + e^{-2s})ds$$

$$= -\frac{1}{16} + \frac{1}{16}\cosh 2t + \frac{1}{16}\sinh 2t + \frac{t}{8}e^{-2t} + \frac{3}{10}\sinh 2t + \frac{3}{20}\sin t.$$

Plugging this last equality into (6.90), we obtain

$$y(t) = \frac{13}{16} - \frac{t^2}{8} + \frac{1}{8}(t+1)e^{-2t} - \frac{3}{5}\sin t + \frac{1}{16}\sinh 2t + \frac{1}{16}\cosh 2t + \frac{4}{5}\sinh 2t$$

as the solution of (6.86).

ℹ Remark 6.2.18 As a side note, we see that equation (6.86) can be also solved by taking the change of variables $u(t) = y'(t)$, then we get a second order equation of $u(t)$, where its solution can be found using the superposition principle together with the method of undetermined coefficients. Once we found $u(t)$, we may integrate it to get $y(t)$. But our goal here is to show how we can use the method discussed in ▶ Sect. 6.2.7 to solve higher-order equations. In addition the change of variables works only in some very particular cases, but the method that we used here works in the general situation.

6.3 Exercises

Exercise 6.1
We consider the system of differential equation

$$Y'(t) = A(t)Y(t), \tag{6.91}$$

with

$$Y(t) = \begin{bmatrix} x(t) \\ y(t) \end{bmatrix}, \quad \text{and} \quad A(t) = \begin{bmatrix} 2/t & -1/t^2 \\ 2 & 0 \end{bmatrix}$$

for $t > 0$.

1. Show that

$$M(t) = \begin{bmatrix} 1/2 & t \\ t & t^2 \end{bmatrix}$$

is a fundamental matrix of system (6.91).

2. Find the resolvent matrix $R(t, s)$ and the solution of the initial value problem

$$\begin{cases} Y'(t) = A(t)Y(t) + B(t), \\ Y(1) = Y_0, \end{cases} \tag{6.92}$$

with

$$B(t) = \begin{bmatrix} t \\ t^2 \end{bmatrix}, \quad \text{and} \quad Y_0 = \begin{bmatrix} 0 \\ 0 \end{bmatrix}$$

Solution

1. In order to show that $M(t)$ is a fundamental matrix of system (6.91), and according to Definition 6.2.3, we need just to prove that the set of the two vectors

$$Y_1(t) = \begin{bmatrix} 1/2 \\ t \end{bmatrix}, \quad \text{and} \quad Y_2(t) = \begin{bmatrix} t \\ t^2 \end{bmatrix}$$

is a fundamental set of solutions of (6.91). Indeed, applying Definition 6.2.2, we deduce that the Wronskian of $Y_1(t)$ and $Y_2(t)$ is

$$W(t) = \det \begin{bmatrix} 1/2 & t \\ t & t^2 \end{bmatrix} = -t^2/2 \neq 0$$

since $t > 0$. Therefore, we can deduce that $Y_1(t)$ and $Y_2(t)$ are two independent solutions of (6.91). Hence, $M(t)$ is a fundamental matrix of (6.91).

2. To find the resolvent matrix $R(t, s)$, we need to find first the inverse $M^{-1}(t)$. Indeed, let $M^{-1}(t)$ be the matrix

$$M^{-1}(t) = \begin{bmatrix} a & b \\ c & d \end{bmatrix}.$$

Then, using the identity $M(t)M^{-1}(t) = I$, we get

$$\begin{cases} a/2 + ct = 1, \\ at + ct^2 = 0, \\ b/2 + dt = 0, \\ bt + dt^2 = 1. \end{cases}$$

Solving the above algebraic system, we find that $a = -2$, $b = 2/t$, $c = 2/t$ and $d = -1/t^2$. Consequently, we get

$$M^{-1}(t) = \begin{bmatrix} -2 & 2/t \\ 2/t & -1/t^2 \end{bmatrix}.$$

Now, applying (6.37), we obtain

$$R(t,s) = \begin{bmatrix} -1 + 2t/s & 1/s - t/s^2 \\ -2t + 2t^2/s & 2t/s - t^2/s^2 \end{bmatrix}.$$

Next, to find the solution of the initial value problem (6.92), we need to apply formula (6.55), to get

$$Y(t) = R(t,1)Y_0 + \int_1^t R(t,s)B(s)ds$$

$$= \int_1^t \begin{bmatrix} t \\ t^2 \end{bmatrix} ds = (t-1) \begin{bmatrix} t \\ t^2 \end{bmatrix}.$$

The following exercise has been proved as a theorem in [13].

Exercise 6.2

Let $A(t)$ be a continuous matrix on an interval J. Show that the following are equivalent for all s, r and t in J:

(i) $A(s)A(t) = A(t)A(s)$,[2]

(ii) $\left(\int_s^r A(v)dv \right) \cdot \left(\int_s^t A(u)du \right) = \left(\int_s^t A(u)du \right) \cdot \left(\int_s^r A(v)dv \right)$,

(iii) $A(r) \int_s^t A(u)du = \int_s^t A(u)du\, A(r)$.

Solution

First, we need to show that (i) implies (ii). Indeed, if (i) holds, then, we have

$$\left(\int_s^r A(v)dv \right) \cdot \left(\int_s^t A(u)du \right) = \int_s^r \int_s^t A(v)A(u)dvdu$$

$$= \int_s^t \int_s^r A(u)A(u)dudv$$

$$= \left(\int_s^t A(u)du \right) \cdot \left(\int_s^r A(v)dv \right).$$

Next, we want to show that (ii) implies (iii). Thus, we have in one hand

$$\frac{d}{dr}\left\{ \left(\int_s^r A(v)dv \right) \cdot \left(\int_s^t A(u)du \right) \right\} = A(r) \int_s^t A(u)du. \tag{6.93}$$

On the other hand, we have

$$\frac{d}{dr}\left\{ \left(\int_s^t A(u)du \right) \cdot \left(\int_s^r A(v)dv \right) \right\} = \int_s^t A(u)du\, A(r). \tag{6.94}$$

Since according to (ii) the two left-hand sides in (6.93) and (6.94) are equal, then the right-hand sides are also identical. This shows (iii).

---

[2] Any matrix $A(t)$ satisfying this assumption is called semiproper matrix.

Finally, we need to prove that (iii) also implies (i). This can be easily shown by computing

$$\frac{d}{dt}\left\{ A(r) \int_s^t A(u)du \right\} = A(r)A(t).$$

Also,

$$\frac{d}{dt}\left\{ \int_s^t A(u)du\, A(r) \right\} = A(t)A(r).$$

This shows that (iii) yields (i).

Exercise 6.3 (Non-constant coefficients)

We consider the system

$$Y'(t) = A(t)Y(t), \tag{6.95}$$

where $A(t)$ is an $n \times n$ continuous matrix on some interval J.

1. Show that if

$$A(s)A(t) = A(t)A(s), \tag{6.96}$$

for all s, t in J then, the resolvent matrix associated to (6.95) is given by

$$R(t, s) = \exp\left\{ \int_s^t A(u)du \right\}. \tag{6.97}$$

2. Find the explicit formula of the resolvent matrix $R(t, s)$ if

$$A(t) = \alpha(t)A + \beta(t)B, \tag{6.98}$$

where $\alpha(t)$ and $\beta(t)$ are two continuous functions on J and A and B are two constant matrices satisfying $AB = BA$.

Solution

1. It is enough to show that

$$M(t) = \exp\left\{ \int_s^t A(u)du \right\}$$

is a solution of the differential equation

$$M'(t) = A(t)M(t), \qquad M(s) = I, \tag{6.99}$$

where I is the identity $n \times n$ matrix. Then, since $R(t, s)$ is the unique solution of (6.99), we deduce that

$$R(t, s) = M(t)$$

$$= \exp\left\{ \int_s^t A(u)du \right\}.$$

Indeed, it is clear that

$$M(s) = \exp\left\{ \int_s^s A(u)du \right\} = I.$$

On the other hand, it is not hard to see that

$$\frac{d}{dt}M(t) = A(t)\exp\left\{ \int_s^t A(u)du \right\}$$

$$= \exp\left\{ \int_s^t A(u)du \right\} A(t)$$

$$= M(t)A(t),$$

where we have used the third property (iii) in Exercise 6.2.

2. It is clear that (6.98) implies (6.96). Therefore from the result of the first question, we get

$$R(t,s) = \exp\left\{ \int_s^t (\alpha(u)A + \beta(u)B)du \right\}$$

$$= \exp\left\{ \int_s^t \alpha(u)du.A \right\} \exp\left\{ \int_s^t \beta(u)du.B \right\}.$$

Exercise 6.4

Find the solution of the system

$$\begin{cases} Y'(t) = A(t)Y(t), \\ Y(0) = Y_0, \end{cases} \tag{6.100}$$

where

$$A(t) = \begin{bmatrix} 0 & -t \\ -t & 0 \end{bmatrix} \quad \text{and} \quad Y_0 = \begin{bmatrix} 4 \\ 2 \end{bmatrix}.$$

Solution

It is clear that

$$A(s)A(t) = \begin{bmatrix} st & 0 \\ 0 & st \end{bmatrix} = A(t)A(s).$$

Thus, we easily apply the result in Exercise 6.3, to get

$$R(t,0) = \exp\left\{ \int_0^t A(u)du \right\} = \exp\left\{ \begin{bmatrix} 0 & -t^2/2 \\ -t^2/2 & 0 \end{bmatrix} \right\}.$$

On the other hand, we have

$$A = -t^2/2 \begin{bmatrix} 0 & 1 \\ 1 & 0 \end{bmatrix} = -\frac{t^2}{2}B,$$

where

$$B = \begin{bmatrix} 0 & 1 \\ 1 & 0 \end{bmatrix}.$$

As we have seen in Example 6.10, we have

$$B^{2k} = I \quad \text{and} \quad B^{2k+1} = B.$$

Therefore,

$$
\begin{aligned}
e^A &= I + A + \frac{A^2}{2!} + \ldots + \frac{A^k}{k!} + \ldots \\
&= I - \frac{t^2}{2}B + \frac{t^4}{4}\frac{B^2}{2!} + \ldots + (-1)^k \frac{t^{2k}}{2^k}\frac{B^k}{k!} + \ldots \\
&= \begin{bmatrix} \cosh(t^2/2) & -\sinh(t^2/2) \\ -\sinh(t^2/2) & \cosh(t^2/2) \end{bmatrix}.
\end{aligned}
$$

Consequently, the solution of (6.100) is

$$Y(t) = R(t,0)Y_0 = \begin{bmatrix} 4\cosh(t^2/2) - 2\sinh(t^2/2) \\ 2\cosh(t^2/2) - 4\sinh(t^2/2) \end{bmatrix}.$$

Exercise 6.5

Solve the system of differential equations

$$\begin{cases} x'(t) = \frac{1}{t}x(t) + t y(t), \\ y'(t) = y(t), \end{cases} \tag{6.101}$$

for $t > 0$ and find the resolvent matrix $R(t,s)$.

Solution

Here, we cannot use formula (6.97), since (6.96) is not satisfied. Instead, we see from (6.101) that the second equation is decoupled from the first one and we can find its solution using the method of separable variables as

$$\frac{dy(t)}{dt} = y(t).$$

This simply gives

$$y(t) = c_1 e^t, \tag{6.102}$$

where c_1 is a constant. Plugging (6.102) into the first equation in (6.101), we obtain

$$\frac{dx(t)}{dt} = \frac{1}{t}x(t) + c_1 t e^t. \tag{6.103}$$

Equation (6.103) is a linear first order equation, then we can easily solve it by using the method of integrating factor introduced in ▶ Sect. 2.3. Indeed, let

$$
\begin{aligned}
\mu(t) &= \exp\left\{\int \frac{-1}{t}\,dt\right\} \\
&= \frac{1}{t}.
\end{aligned}
$$

Now, multiplying equation (6.103) by $\mu(t) = 1/t$, we obtain

$$\frac{d}{dt}\left\{\frac{1}{t}x(t)\right\} = c_1 e^t.$$

Integrating both sides in the above equation and using integration by parts, we obtain

$$\frac{1}{t}x(t) = c_1 e^t + c_0.$$

Thus,

$$x(t) = c_1 t e^t + c_0 t.$$

Consequently, the solution of (6.101) is

$$Y(t) = M(t)C,$$

where

$$Y(t) = \begin{bmatrix} x(t) \\ y(t) \end{bmatrix}, \quad M(t) = \begin{bmatrix} t & te^t \\ 0 & e^t \end{bmatrix} \quad \text{and} \quad C = \begin{bmatrix} c_0 \\ c_1 \end{bmatrix}.$$

It is not hard to verify that

$$M^{-1}(t) = \begin{bmatrix} 1/t & -1 \\ 0 & e^{-t} \end{bmatrix}.$$

Thus,

$$R(t,s) = M(t)M^{-1}(s) = \begin{bmatrix} t/s & -t + te^{t-s} \\ 0 & e^{t-s} \end{bmatrix}.$$

Exercise 6.6
We consider the initial value problem

$$\begin{cases} \dfrac{d^n y(t)}{dt^n} = b(t), \\ y(t_0) = y'(t_0) = \ldots = y^{n-1}(t_0) = 0, \end{cases} \tag{6.104}$$

where $b(t)$ is a continuous function on some interval J that contains t_0. Show that for all t, then the solution of (6.104) is

$$y(t) = \int_{t_0}^{t} \frac{(t-s)^{n-1}}{(n-1)!} b(s)ds. \tag{6.105}$$

Solution
Without loss of generality, we may take $t_0 = 0$. By using the change of variables

$$y_1 = y, \quad y_2 = y', \quad \ldots, \quad y_n = y^{(n-1)},$$

then, equation (6.104) can be written as a first order system of the form

$$Y'(t) = AY(t) + B(t) \tag{6.106}$$

with

$$Y(t) = \begin{bmatrix} y_1(t) \\ y_2(t) \\ \vdots \\ y_n(t) \end{bmatrix}, \qquad B(t) = \begin{bmatrix} 0 \\ 0 \\ \vdots \\ b(t) \end{bmatrix}$$

and A is the matrix

$$A = \begin{bmatrix} 0 & 1 & 0 & \cdots & 0 \\ 0 & 0 & 1 & \cdots & 0 \\ \vdots & \vdots & & \vdots & \vdots \\ 0 & 0 & \cdots & 0 & 1 \\ 0 & 0 & \cdots & \cdots & 0 \end{bmatrix}.$$

Our goal now is to use formula (6.73) to compute e^{tA}. Thus, since $\lambda_0 = 0$ is the only eigenvalue of A with algebraic multiplicity n, then it is easy to find the functions r_j, $1 \le j \le n$ in (6.74) as

$$\begin{cases} r_1(t) = 1, \\ r_2(t) = t, \\ \vdots \\ r_n = \dfrac{t^{n-1}}{(n-1)!}. \end{cases} \tag{6.107}$$

On the other hand, we may compute the matrices P_j in Theorem 6.2.16 as

$$P_0 = I, \quad P_1 = A, \ldots, P_{n-1} = A^{n-1}.$$

Consequently, formula (6.73) leads to

$$e^{tA} = I + tA + \ldots + \frac{t^{n-1}}{(n-1)!} A^{n-1}.$$

A simple computation gives

$$e^{tA} = \begin{bmatrix} 1 & t & \frac{1}{2}t^2 & \cdots & \frac{1}{(n-1)!}t^{n-1} \\ 0 & 1 & t & \cdots & \frac{1}{(n-2)!}t^{n-2} \\ \vdots & \vdots & & \vdots & \vdots \\ 0 & 0 & \cdots & 0 & t \\ 0 & 0 & \cdots & \cdots & 1 \end{bmatrix}.$$

Using (6.55), we deduce that the solution of (6.104) is the first component of the vector

$$Y(t) = \int_0^t e^{(t-s)A} B(s)\,ds$$

$$= \int_0^t \frac{(t-s)^{n-1}}{(n-1)!} b(s)\,ds.$$

Qualitative Theory

Belkacem Said-Houari

B. Said-Houari, *Differential Equations: Methods and Applications,* Compact Textbooks in Mathematics,
DOI 10.1007/978-3-319-25735-8_7, © Springer International Publishing Switzerland 2015

We have seen in the previous chapters how to solve some differential equations using different methods. However, due to the lack of a general method of solving general differential equations, we do not know the explicit solutions of many important differential equations or sometimes we need just to know some information about the solutions like their behavior when t (the independent variable) becomes very large. Furthermore, maybe it is not worth it to find these solutions just in order to know some little information about them. This chapter will be devoted to the discussion of how solutions of certain differential equations or systems of differential equations behave. This knowledge must rely only on the properties of the vector fields that define the differential equations. Therefore, our goal here is to describe the solutions rather than finding them. Especially, this approach will help when dealing with nonlinear problems, since many physical systems can be described by nonlinear differential equations. This is known as the *qualitative* study of differential equations and it is one of the major aspects and the biggest achievement of the modern analysis of differential equations. This theory was mainly developed by H. Poincaré and A.M. Lyapunov. Since the qualitative theory of differential equations is quite difficult, so, in most cases, we limit our discussion to the two dimensional systems.

7.1 Two-Dimensional Autonomous Systems and Phase Space

In this section, we consider the planar system of the following form

$$\begin{cases} \dfrac{dx}{dt} = F(x, y), \\ \dfrac{dy}{dt} = G(x, y), \end{cases} \tag{7.1}$$

where $x = x(t)$, $y = y(t)$ and $F(x, y)$ and $G(x, y)$ are continuous and have continuous first partial derivatives in some domain Ω in the xy-plane. System (7.1) in which the independent variable t does not appear explicitly in the functions $F(x, y)$ and $G(x, y)$ is called *autonomous* systems. As we will see later, the solution of (7.1) can be represented as a curve in the xy-plane called the *phase plane* with t as a parameter.

Now, we study the solutions of such system.

> **Definition 7.1.1 (Critical point)**
>
> Any point (x_0, y_0) in Ω satisfying $F(x_0, y_0) = G(x_0, y_0) = 0$ is called a *critical point* of (7.1). Any other point in Ω is called *regular*.

The critical points are also called singular points and points of equilibrium.

Example 7.1

Find the critical points of the system

$$\begin{cases} \dfrac{dx}{dt} = y^2 - 5x + 6, \\ \dfrac{dy}{dt} = x - y. \end{cases} \tag{7.2}$$

Solution

In order to find the critical points of (7.2), we need to solve the system of equations

$$\begin{cases} y^2 - 5x + 6 = 0, \\ x - y = 0. \end{cases}$$

The solution of the above system gives $x = y = 2$ or $x = y = 3$. Thus system (7.2) has two critical points $(2, 2)$ and $(3, 3)$.

> **Lemma 7.1.1** Let $x = x(t)$ and $y = y(t)$ be a solution of (7.1) for $a < t < b$, then for any constant c,
>
> $$x_1(t) = x(t + c), \qquad y_1(t) = y(t + c)$$
>
> is also a solution of (7.1) in the interval $a - c < t < b - c$.

Proof

Using the chain rule, we have

$$\frac{dx_1(t)}{dt} = \frac{dx(t + c)}{dt} = F(x(t + c), y(t + c))$$
$$= F(x_1(t), y_1(t)).$$

and similarly

$$\frac{dy_1(t)}{dt} = \frac{dy(t + c)}{dt} = G(x(t + c), y(t + c))$$
$$= G(x_1(t), y_1(t)),$$

which are defined for $a - c < t < b - c$.

It is clear that if (x_0, y_0) is a critical point, then system (7.1) has the constant solution

$$
\begin{bmatrix} x(t) \\ y(t) \end{bmatrix} = \begin{bmatrix} x_0 \\ y_0 \end{bmatrix}.
$$

The vector $V(x, y) = (F(x, y), G(x, y))$ defines a vector field in Ω and a solution $(x(t), y(t))$ of system (7.1) can be interpreted as the resulting path or trajectory or orbit of a particular as it moves through the region Ω where its velocity at each point (x, y) in Ω is the the vector $V(x, y)$. The region Ω is called the *phase space* of system (7.1).

Definition 7.1.2 (Isolated critical point)

A critical point (x_0, y_0) of (7.1) is called an *isolated* critical point if there exists a neighborhood of (x_0, y_0) containing no other critical points.

In the sequel, we denote the distance between two points $X = (x_1, x_2)$ and $Y = (y_1, y_2)$ by $\|X - Y\|$ defined by

$$
\|X - Y\| = \sqrt{(x_1 - y_1)^2 + (x_2 - y_2)^2}.
$$

Definition 7.1.3 (Stability of critical point)

Let $X_1 = (x_1, y_1)$ be an isolated critical point of (7.1) and let $X(t) = (x(t), y(t))$ be the solution of (7.1) that satisfies the initial condition $X(t_0) = X_0$ for some t_0 where $X_0 \neq X_1$.
- We say that X_1 is *stable* when given any $\rho > 0$, there is a corresponding $r > 0$ such that if the initial position X_0 satisfies $\|X_0 - X_1\| < r$, then the corresponding solution $X(t)$ satisfies $\|X(t) - X_1\| < \rho$ for all $t \geq t_0$.
- We say that X_1 is *asymptotically stable* if in addition $\lim_{t \to \infty} X(t) = X_1$ whenever $\|X_0 - X_1\| < r$.
- If X_1 is not stable is said to be *unstable*.

Example 7.2
We consider the system

$$
\begin{cases} \dfrac{dx}{dt} = -x, \\ \dfrac{dy}{dt} = -y, \end{cases} \tag{7.3}
$$

with $x(0) = y(0) = 1$. Show that the point (0.0) is asymptotically stable.

Solution
It is clear that the point $X_1 = (0, 0)$ is an isolated critical point. The solution of (7.3) is $X(t) = (x(t), y(t))$ with $x(t) = y(t) = e^{-t}$. Let $X(0) = (1, 1)$ and let $\rho > 0$. Then we can find $r = \rho$ such that if $\|X_0 - X_1\| = \sqrt{2} < \rho$, then $\|X(t) - X_1\| = \sqrt{2}e^{-t} < \sqrt{2} < \rho$ for all $t \geq 0$. This shows that X_1 is stable. Since

$$
\lim_{t \to \infty} X(t) = \lim_{t \to \infty} (e^{-t}, e^{-t}) = X_1.
$$

Then X_1 is asymptotically stable.

7.1.1 Stability of Linear Systems

In this subsection, we consider the particular case of system (7.1) where $F(x, y)$ and $G(x, y)$ are linear functions of x and y. Thus, we consider the system

$$\begin{cases} \dfrac{dx}{dt} = ax + by, \\[2mm] \dfrac{dy}{dt} = cx + dy. \end{cases} \qquad (7.4)$$

There are four simple types of critical points that occur quite frequently:
1. Nodes;
2. Saddle points;
3. Centers;
4. Spirals.

We give a definition to the above critical points.

┌─ **Definition 7.1.4** ─────────────────────────────────────

Assume that (x_0, y_0) is an isolated critical point of system (7.1).
- If all the trajectories near the critical point may approach the critical point along curves which are asymptotically straight lines as $t \to \infty$, then we call such critical point a *stable node*. On the other hand if all the trajectories that start near the critical point move away from it, as t increases, along paths that are approximately straight lines, at least until the trajectory gets away from the critical point, then this critical point is called an *unstable node*.
- If some trajectories may approach the critical point while others move away from the critical point as $t \to \infty$, then such critical point is called a *saddle point*.
- If all the trajectories may form closed orbits about the critical point, then such critical point is called a *center*.
- If all trajectories may approach the critical point along spiral curves as $t \to \infty$, then such critical point is called a *stable spiral point*. On the other hand, if all trajectories, when t increases, move away from the critical point along paths that are, at least initially, spiral shaped is called *unstable spiral point*.

└──

Now, we discuss the above possibilities for the linear system (7.4). First, it is clear that $(x_0, y_0) = (0,0)$ is a critical point of (7.4) and we assume that there are no other critical points. This is equivalent to assume that $ad - bc \neq 0$.

We may write system (7.4) in the matrix form as

$$\frac{dX(t)}{dt} = AX(t), \qquad (7.5)$$

where

$$X(t) = \begin{bmatrix} x(t) \\ y(t) \end{bmatrix} \qquad \text{and} \qquad A = \begin{bmatrix} a & b \\ c & d \end{bmatrix}.$$

It is easy to classify this critical point once the eigenvalues of the matrix A are known. According to what we have seen in ▶ Chapter 6, if λ_1 and λ_2 are two distinct real solutions of the

■ **Fig. 7.1** Stable node

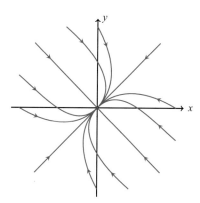

■ **Fig. 7.2** Unstable node

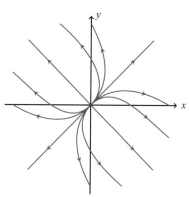

characteristic equation

$$\lambda^2 - (a+d)\lambda + (ad - bc) = 0,$$

and V_1 and V_2 are the corresponding eigenvectors, respectively, then the solution of (7.5) is given by

$$X(t) = c_1 V_1 e^{\lambda_1 t} + c_2 V_2 e^{\lambda_2 t},$$

where c_1 and c_2 are two constants which can be determined by the initial position $X(0)$. We distinguish the following important cases:

Case 1: The two eigenvalues are real and negative. It is clear that if λ_1 and λ_2 are both negative, then all the trajectories approaches the critical point $(0, 0)$ and the critical point is a *stable node* as shown in ■ Fig. 7.1.

Case 2: The two eigenvalues are real and positive. If λ_1 and λ_2 are both real and positive, then all the trajectories move away from $(0, 0)$ as $t \to \infty$ and $(0, 0)$ is an *unstable node* as in ■ Fig. 7.2.

Case 3: One eigenvalue is positive and the other is negative. If λ_1 is negative and λ_2 is positive, then $(0, 0)$ is a *saddle point*; the trajectories approach the origin in the direction of V_1 and move away from the origin in the direction of V_2 as in ■ Fig. 7.3.

Case 4: The eigenvalues are complex conjugate. If the eigenvalues of the matrix A are complex, then they must be conjugate since A is a real matrix, that is $\lambda_1 = \alpha + i\beta$ and

Fig. 7.3 Saddle point

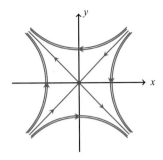

Fig. 7.4 Center (stable)

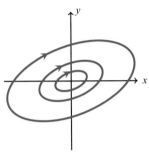

$\lambda_2 = \alpha - i\beta$ where α and β are real numbers. If $\alpha = 0$, then the two eigenvalues are pure imaginary and then $X(t)$ represents a closed orbit for any c_1 and c_2 and therefore the critical point is a *center* as in ☐ Fig. 7.4. On the other hand if $\alpha \neq 0$ then the critical point is a *spiral* which is stable if $\alpha < 0$ and unstable if $\alpha > 0$ as in ☐ Fig. 7.5 and ☐ Fig. 7.6, respectively.

Fig. 7.5 Spiral (stable)

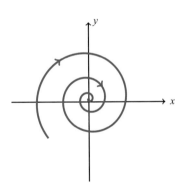

Fig. 7.6 Spiral (unstable)

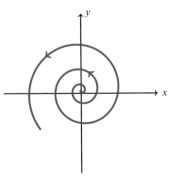

Example 7.3

We consider the system of differential equations

$$\begin{cases} \dfrac{dx}{dt} = -2x + y, \\ \dfrac{dy}{dt} = x - 2y, \end{cases} \tag{7.6}$$

Find the nature of the critical point $(0, 0)$.

Solution

It is not hard to verify that the eigenvalues of the matrix

$$A = \begin{bmatrix} -2 & 1 \\ 1 & -2 \end{bmatrix}$$

are $\lambda_1 = -3$ and $\lambda_2 = -1$ and the corresponding eigenvectors are, respectively,

$$V_1 = \begin{bmatrix} 1 \\ -1 \end{bmatrix} \quad \text{and} \quad V_2 = \begin{bmatrix} 1 \\ 1 \end{bmatrix}.$$

Thus, the solution of (7.6) is

$$X(t) = \begin{bmatrix} x(t) \\ y(t) \end{bmatrix} = c_1 V_1 e^{-3t} + c_2 V_2 e^{-t},$$

where c_1 and c_2 are two constants. Consequently, the critical point $(0, 0)$ is a node that is asymptotically stable.

We want to mention that in the case of a center or spiral and in order to determine whether the rotation is clockwise or counterclockwise, we simply need to pick a point $(\hat{x}, \hat{y})$ in the orbit and then the direction of the velocity vector $V = (F(\hat{x}, \hat{y}), G(\hat{x}, \hat{y}))$ at this point will give us the direction of the rotation.

Before going to the general theory, let us first give a simple physical model of a nonlinear system of differential equations.

7.1.2 The Two Dimensional Nonlinear System

It has been known that it is difficult to find the explicit solution of nonlinear systems of differential equations, so it is reasonable to look for some qualitative properties of the solutions, thus the critical point analysis is mostly used to find the global features of some nonlinear systems of equations. Because of the powerful tools we know for linear systems, the first step to analyze a nonlinear system is usually to linearize it, if possible. Although this method has its limitations, it still give some information, at least about the local behavior near the critical points of the nonlinear system.

The Motion of a Pendulum One of the most interesting physical example modeled by system (7.1) is the nonlinear damped pendulum. Consider a simple pendulum as shown in ◼ Fig. 7.7,

□ **Fig. 7.7** Pendulum

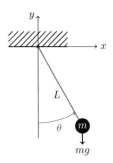

where L denotes the length of the rod and m denotes the mass of the bob. Assume that the rod is rigid and has zero mass. Let θ denote the angle subtended by the rod and the vertical axis through the pivot point and considered positive when measured to the right of the vertical axis and negative to the left. The pendulum is free to swing in the vertical plane about the equilibrium position. In order to write the differential equation that governs the motion of the pendulum, we first describe the forces acting on the bob as:

- a downward gravitational force equals mg, where g is the gravity;
- the frictional force resisting the motion which assumes to be proportional to the speed of the bob with a friction coefficient b.

The equation of motion of the pendulum in the tangential direction can be written by applying Newton's second law, in the direction perpendicular to the rod, to give

$$mL\theta''(t) = -mg \sin \theta - bL\theta',$$

where $L\theta''$ is the tangential acceleration, $mg \sin \theta$ is the tangential gravity force (the minus indicates the restoring nature of the force) and $L\theta'$ is the tangential velocity. The above equation can be written as

$$\theta''(t) + \gamma\theta' + \eta \sin \theta = 0 \qquad (7.7)$$

with $\gamma = b/m$ and $\eta = g/L$. Introducing the variables $x = \theta$ and $y = \theta'$, we obtain the system

$$\begin{cases} \dfrac{dx}{dt} = y, \\[2mm] \dfrac{dy}{dt} = -\eta \sin x - \gamma y. \end{cases} \qquad (7.8)$$

Thus, system (7.8) is written in the general form (7.1) with $F(x, y) = y$ and $G(x, y) = \eta \sin x - \gamma y$. To find the critical points of system (7.8), we need to solve simultaneously the equations

$$F(x, y) = G(x, y) = 0.$$

Then, we get the critical points of the form $(n\pi, 0)$ where $n = 0, \pm 1, \pm 2, \dots$ But physically there are only two critical points $(0, 0)$ and $(\pi, 0)$ and all the other critical points are repetitions of these two points.

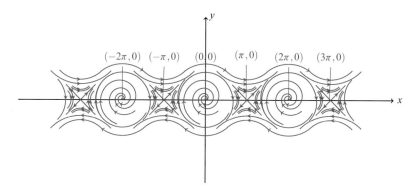

◻ Fig. 7.8 The big picture of the phase space of the nonlinear pendulum. We see that while the pendulum can rest at the critical points $(2m\pi, 0)$, $m = 0, \pm1, \pm2, \ldots$, it can hardly maintain rest at the critical points $((2m + 1)\pi, 0)$ because infinitesimally small disturbance from those critical points will take the pendulum away

Now, we suppose that the pendulum is slightly damped, by assuming that

$$\gamma^2 < 4\eta. \tag{7.9}$$

By linearizing system (7.8) near the critical point $(0, 0)$ and for small values of x, we approximate $\sin x$ by x, then, we get the linearized system

$$\begin{cases} \dfrac{dx}{dt} = y, \\ \dfrac{dy}{dt} = -\eta x - \gamma y. \end{cases} \tag{7.10}$$

The eigenvalues of the matrix

$$A = \begin{bmatrix} 0 & 1 \\ -\eta & -\gamma \end{bmatrix}$$

are

$$\lambda_1 = \frac{-\gamma + \sqrt{\gamma^2 - 4\eta}}{2}, \qquad \lambda_2 = \frac{-\gamma - \sqrt{\gamma^2 - 4\eta}}{2}.$$

It is clear that under the assumption (7.9), the eigenvalues are complex and thus, the critical point $(0, 0)$ is a stable spiral since the real parts of the eigenvalues are negative. To find the direction of the rotation, we simply see that the vector

$$V = \begin{bmatrix} 0 \\ -\eta \end{bmatrix}$$

is a vector from the velocity field and since $\eta > 0$, then the direction of the rotation is clockwise as in ◻ Fig. 7.8.

Next, for the critical point $(\pi, 0)$, we linearize the system as follows, we compute the matrix of the linearized system to be the Jacobian at $(\pi, 0)$. Thus,

$$A = J_{(\pi,0)} = \begin{bmatrix} F_x(\pi,0) & F_y(\pi,0) \\ G_x(\pi,0) & G_y(\pi,0) \end{bmatrix}$$

$$= \begin{bmatrix} 0 & 1 \\ \eta & -\gamma \end{bmatrix}.$$

Now, the eigenvalues of the matrix A are

$$\hat{\lambda}_1 = \frac{-\gamma + \sqrt{\gamma^2 + 4\eta}}{2}, \qquad \hat{\lambda}_2 = \frac{-\gamma - \sqrt{\gamma^2 + 4\eta}}{2}.$$

Hence, the above eigenvalues are real with $\lambda_1 > 0$ and $\lambda_2 < 0$. Therefore, the critical point $(\pi, 0)$ is a saddle point as in ◼ Fig. 7.8.

Now, coming back to the general system (7.1) and in order to plot the trajectories near the critical points, we have to do the following:

1. Find the critical points, by solving simultaneously the algebraic equations $F(x, y) = 0$ and $G(x, y) = 0$.
2. For each critical point, we linearize the system near the critical point by computing the Jacobian matrix at the critical point.
3. Find the nature of the critical points of the linearized system and plot the trajectories. Then, add all the other trajectories which are compatible with the one of the linearized system as we did for the nonlinear pendulum.

7.1.3 Limit Cycles

From the above discussion, we can know the path of (7.1) only in the neighborhood of certain type of critical points and as we have seen, the phase space of system (7.1) can only consist of critical points, nonintersecting trajectories or closed curves called *cycles*. In this subsection we are interested on the global properties of (7.1), in particular, we are going to investigate some closed paths or trajectories called *limit cycles*.

┌─ Definition 7.1.5 ───────────────────────────────────

Let K be a cycle of the phase space of (7.1). Then, K is called a *limit cycle* if there exists a neighborhood of K such that any trajectory passing through the neighborhood is not a cycle.

So, a limit cycle is an isolated closed trajectory in the phase plane and having one of the following properties:

- Trajectories near K spiral toward K and approach the limit cycle as time goes to infinity. In this case the limit cycle is called *stable* or *attractive* limit cycle (ω-limit cycle), as in ◼ Fig. 7.9.
- Trajectories near K spiral away from K, as time goes to infinity. In this case the limit cycle is called *unstable* or (a-limit cycle).
- Some trajectories near K approaches K and other trajectories go away from K, in this case the limit cycle is called *semistable*.

□ Fig. 7.9 Stable Limit cycle

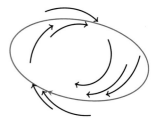

There are limit cycles which are neither stable, unstable nor semistable. Here we are interested mainly on the stable limit cycles and in the sequel, we mean by limit cycle a stable limit cycle. If a system has a stable limit cycle and we disturb it, then after some time it will go eventually to this limit. In general it is quit hard to tell in advance whether a given system of differential equations has a limit cycle among its solutions, but such limit cycles are known to exists only in nonlinear dissipative systems.

Nonexistence of Limit Cycles Now, let us consider system (7.1) and the question that we should ask is the following: *under which conditions on the functions $F(x, y)$ and $G(x, y)$ system (7.1) has stable limit cycles? And if so, how many?* The answer to this question is not fully well known. In the next theorem, we give the Bendixson criterion of the nonexistence of limit cycles.

Theorem 7.1.2 (Bendixson's negative criterion)

Let D be a simply connected region in the (x, y)-phase plane. Assume that $\frac{\partial F}{\partial x}$ and $\frac{\partial G}{\partial y}$ exist and are continuous functions on D. If the divergence of the vector field $V(x, y) = (F(x, y), G(x, y))$ is not zero, that is if

$$\text{div } V(x, y) = \frac{\partial F}{\partial x}(x, y) + \frac{\partial G}{\partial y}(x, y) \neq 0,$$

then there is no closed trajectory of the system (7.1) in the region D and therefore there is no limit cycle of this system in D.

Proof

We assume that system (7.1) has a closed orbit K and we reach to a contradiction. Let R be the region enclosed by K. For the vector field $V(x, y) = (F(x, y), G(x, y))$, Green's theorem implies

$$\oint_K V \cdot \mathbf{n} ds = \iint_R \text{div } V(x, y) dA, \tag{7.11}$$

where $\mathbf{n}$ is the unit normal vector which is perpendicular to V. Thus, the dot product $V \cdot \mathbf{n} = 0$. Therefore the left-hand side in (7.11) is identically zero. On the other hand, since **div** $V(x, y) \neq 0$ on D, then **div** $V(x, y) \neq 0$ on R and therefore **div** $V(x, y) < 0$ everywhere in R or **div** $V(x, y) > 0$ everywhere in R. Thus, the integral in the right-hand side of (7.11) must be either strictly positive or strictly negative, but not zero. This is a contradiction, since we have already shown that the left-hand side of (7.11) is zero.

Example 7.4

Show that the damped nonlinear pendulum described by system (7.8), that is

$$\begin{cases} \dfrac{dx}{dt} = y, \\[2mm] \dfrac{dy}{dt} = -\eta \sin x - \gamma y. \end{cases} \tag{7.12}$$

has no limit cycle if $\gamma \neq 0$.

Solution

We consider the vector field $V(x, y) = (y, -\eta \sin x - \gamma y)$. Then, we have

$$\mathbf{div}\, V(x, y) = \frac{\partial}{\partial x}(y) + \frac{\partial}{\partial y}(-\eta \sin x - \gamma y)$$

$$= -\gamma.$$

Since $\mathbf{div}\, V(x, y) = -\gamma \neq 0$, then according to Theorem 7.1.2, system (7.12) has no limit cycle.

Example 7.5

Does the following system has a limit cycle

$$\begin{cases} \dfrac{dx}{dt} = x^2 + y^2, \\[2mm] \dfrac{dy}{dt} = x^2 - y^2 - 1? \end{cases} \tag{7.13}$$

Solution

The vector field V is defined by $V(x, y) = (x^2 + y^2, x^2 - y^2 - 1)$. Therefore,

$$\mathbf{div}\, V(x, y) = 2x - 2y.$$

It is clear that $\mathbf{div}\, V(x, y) = 0$ along the line $y = x$. Therefore, there is no limit cycle above this line and no limit cycle below this line, but, it may have a limit cycle crossing the line $y = x$. Consequently, in this case Theorem 7.1.2 is not conclusive.

In the following theorem we give another criterion for limit cycles.

Theorem 7.1.3 (Critical-point criterion)
The interior of every periodic orbit of the system (7.1) must contain at least one critical point.

We omit the proof of Theorem 7.1.3. We may also clarify that Theorem 7.1.3 can be used in the following form:

If the system (7.1) has no critical point, then it has no closed trajectory and therefore it has no limit cycle.

Example 7.6

Show that the following system has no limit cycles

$$\begin{cases} \dfrac{dx}{dt} = x^2 + y^2 + 1, \\ \dfrac{dy}{dt} = x^2 - y^2. \end{cases} \qquad (7.14)$$

Solution

It is clear that system (7.14) has no critical point, since $x^2 + y^2 + 1 \neq 0$ for all (x, y) in $\mathbb{R}^2$. Therefore, applying Theorem 7.1.3, we deduce that (7.14) has no limit cycle.

Liénard's System and the Existence of Limit Cycles Our goal now is to give a criterion for the existence of a limit cycle of the system (7.1). There is a particular case of system (7.1) called *Liénard's system* in which the system has a unique limit cycle.

The Liénard equation models the RLC circuit and can be written in the form

$$w''(t) + f(w)w'(t) + g(w) = 0. \qquad (7.15)$$

Using the change of variables

$$\begin{cases} F(w) = \displaystyle\int_0^w f(\zeta)d\zeta, \\ x = w, \\ y = w' + F(w), \end{cases}$$

then equation (7.15) can be transformed into the following system:

$$\begin{cases} \dfrac{dx}{dt} = y - F(x), \\ \dfrac{dy}{dt} = -g(x). \end{cases} \qquad (7.16)$$

In the special case where $F(x) = \mu(\frac{x^3}{3} - x)$ and $g(x) = x$, where μ is a positive physical constant, this system is called the *van der Pol equation*[1] which can be written in the form

$$\frac{d^2x}{dt^2} + \mu(x^2 - 1)\frac{dx}{dt} + x = 0. \qquad (7.17)$$

For $\mu = 0$ or $|x| = 1$, equation (7.17) is simply an equation of harmonic oscillator. On the other hand for $|x| > 1$, then equation (7.17) is damped and for $|x| < 1$, then it gains energy.

Rayleigh's Equation The *Rayleigh equation* has the form

$$x''(t) + \mu\left(\frac{1}{3}(x'(t))^3 - x'(t)\right) + x(t) = 0 \qquad (7.18)$$

---

[1] This equation was introduced by van der Pol in 1926 in a study of the nonlinear vacuum tube circuits of early radios.

where $\mu > 0$. This equation was originated in connection with a theory of the oscillation of a violin string and was derived by Rayleigh in 1877.

The Rayleigh equation can be reduced to the van der Pol equation as follows: differentiating equation (7.18), we get

$$x'''(t) + \mu\left((x'(t))^2 x''(t) - x''(t)\right) + x'(t) = 0.$$

We put $y(t) = x'(t)$, we obtain the van der Pol equation

$$y''(t) + \mu\left((y(t))^2 - 1\right)y'(t) + y(t) = 0. \tag{7.19}$$

It has been known that Rayleigh's equation has a limit cycle.

Theorem 7.1.4 (Liénard's theorem)
We consider system (7.16) and assume that $F(x)$ and $g(x)$ are smooth, odd functions such that $g(x) > 0$ for $x > 0$ and $F(x)$ has exactly three zeros 0, a and $-a$ with $f'(0) < 0$ and $F'(x) \geq 0$ for $x > a$ and $F(x) \to \infty$ for $x \to \infty$. Then system (7.16) has exactly one stable limit cycle.

For the proof of Theorem 7.1.4, the reader is refereed to [17].

Example 7.7
Show that the van der Pol equation (7.17) has one stable limit cycle.

Solution
As we have said before, equation (7.17) can be written in the form of the Liénard system with

$$F(x) = \mu(x^3/3 - x)$$

and

$$g(x) = x.$$

It is clear that both functions $F(x)$ and $g(x)$ are odd functions, $g(x) > 0$ for $x > 0$ and $F(x)$ has three zeros 0, $\sqrt{3}$ and $-\sqrt{3}$. We have also

$$F'(0) = -\mu < 0$$

and since $f'(x) > 0$ for $x > 1$, then $F'(x) > 0$ for $x > \sqrt{3}$, in addition it is clear that $F(x) \to \infty$ for $x \to \infty$. Therefore, Theorem 7.1.4 shows the existence of a unique stable limit cycle of the equation (7.17).

The following theorem has been proved by Zhang in [26] and his result complements Liénard's theorem.

Theorem 7.1.5 (Zhang's theorem)
Consider the Liénard system (7.16) where $F(x)$ and $g(x)$ are continuous and differentiable on $\mathbb{R}$ and let $\alpha < 0 < \beta$. Assume that the following assumptions hold:
(i) $xg(x) > 0$, for $x \neq 0$ and $G(x) \to \infty$ as $x \to \alpha$ if $\alpha = -\infty$ and $G(x) \to \infty$ as $x \to \beta$ if $\beta = \infty$ with $G(x) = \int_0^x g(\zeta)d\zeta$.
(ii) $f(x)/g(x)$ is increasing for x in $(\alpha, 0) \cap (0, \beta)$ and $f(x)/g(x) \neq 0$ in a neighborhood of the origin.
Then, (7.16) has at most one limit cycle. Moreover, the limit cycle is stable if it exists.

Hilbert's 16th Problem Hilbert's 16th problem, or at least the second part of it, asks for the maximal number of limit cycles that system (7.1) can have, when $F(x, y)$ and $G(x, y)$ are polynomials of degree n. This problem was posed by David Hilbert at a conference in Paris in 1900 as a part of his list of 23 problems in mathematics and to my knowledge, this problem has not been yet solved and no upper bound to the number of limit cycles is known. It has been also listed by Smale in 1998 in his book *Mathematical Problems for the Next Century* [23]. He also suggested to better look for the number of limit cycles in the Liénard system with $F(x)$ is a polynomial of degree $2n + 1$ and satisfying the assumption $F(0) = 0$. It is known that the Liénard system has a finite number of limit cycles for each $F(x)$ and the number of limit cycles is the number of fixed point of a mapping called the "Poincaré section". See [23] for more details.

7.2 Exercises

Exercise 7.1
We consider the system of differential equations

$$\begin{cases} \dfrac{dx}{dt} = x, \\ \dfrac{dy}{dt} = y, \end{cases} \tag{7.20}$$

Find the nature of the critical point $(0, 0)$.

Solution
The matrix associated to (7.20) is

$$A = \begin{bmatrix} 1 & 0 \\ 0 & 1 \end{bmatrix}$$

with the eigenvalues $\lambda_1 = \lambda_2 = 1$ and the solution of (7.20) is

$$X(t) = \begin{bmatrix} x(t) \\ y(t) \end{bmatrix} = c_1 V_1 e^t + c_2 V_2 e^t,$$

◻ Fig. 7.10 Unstable node

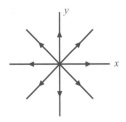

where

$$V_1 = \begin{bmatrix} 1 \\ 0 \end{bmatrix} \quad \text{and} \quad V_2 = \begin{bmatrix} 0 \\ 1 \end{bmatrix}.$$

Since the eigenvalues are equal and positive, then the trajectories move away from $(0,0)$ as $t \to \infty$ and $(0,0)$ is an unstable node as in ◻ Fig. 7.10.

Exercise 7.2 (Lotka–Volterra system)

Find the critical points and their nature in the Lotka–Volterra system

$$\begin{cases} \dfrac{dx}{dt} = ax - bxy, \\ \dfrac{dy}{dt} = -cy + dxy, \end{cases} \tag{7.21}$$

where a, b, c and d are positive constants.

Solution

The critical points of system (7.21) are the solutions to the system

$$\begin{cases} F(x,y) = ax - bxy = 0, \\ G(x,y) = -cy + dxy = 0. \end{cases}$$

The solutions of the above algebraic system are the two critical points $(0,0)$ and $(c/d, a/b)$.
For the first critical point $(0,0)$, the linearized system about this point is

$$\begin{cases} \dfrac{dx}{dt} = ax, \\ \dfrac{dy}{dt} = -cy. \end{cases} \tag{7.22}$$

The matrix associated to (7.22) is

$$A = \begin{bmatrix} a & 0 \\ 0 & -c \end{bmatrix}.$$

The eigenvalues of A are $\lambda_1 = a$ and $\lambda_2 = -c$. Therefore the critical point $(0,0)$ is a saddle point as in ◻ Fig. 7.11 which shows for instance on the x axis that if the number of predators y (foxes) is zero, then the number of preys x (rabbits) grows rapidly since there is no foxes hunting them. On the other hand on the y axis

◻ Fig. 7.11 The point $(0, 0)$ is a saddle point

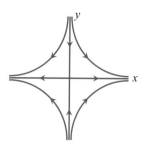

◻ Fig. 7.12 The point $(1, 1)$ is a center

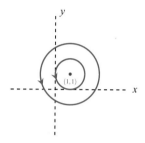

the number of rabbits is zero, then the number of foxes deceases rapidly due to the starvation. The other parts of the phase space shows that if the number foxes deceases, then the number of rabbits grows and vice versa.

For the second critical point $(c/d, a/b)$, we do the linearization by calculating the Jacobian matrix at $(c/d, a/b)$ as

$$A = J_{(c/d, a/b)} = \begin{bmatrix} F_x(c/d, a/b) & F_y(c/d, a/b) \\ G_x(c/d, a/b) & G_y(c/d, a/b) \end{bmatrix}$$

$$= \begin{bmatrix} 0 & -\frac{bc}{d} \\ \frac{ad}{b} & 0 \end{bmatrix}.$$

Without loss of generality, we may take $a = b = c = d = 1$. Then, the matrix A takes the form:

$$A = \begin{bmatrix} 0 & -1 \\ 1 & 0 \end{bmatrix}.$$

Therefore, the eigenvalues of A are complex with real part equals to zero. Thus, the critical point $(1, 1)$ is a center going counterclockwise as in ◻ Fig. 7.12. For the point $(1, 1)$, we cannot deduce anything about the global behavior of the nonlinear system at the critical point $(1, 1)$, since a small perturbation of the linearized system leads to spirals. Therefore the global analysis of the nonlinear system cannot be justified by the local analysis which gives an incomplete description of the nature of the phase-space trajectories.

Now, to know the global behavior of system (7.21), we look for the integral curve of the system by eliminating t and get

$$\frac{dy}{dx} = \frac{y(x-1)}{x(1-y)}. \tag{7.23}$$

Equation (7.23) is a separable equation and can be solved using the method is ▶ Sect. 2.1 to find

$$\frac{(1-y)dy}{y} = \frac{(x-1)dx}{x}.$$

◻ Fig. 7.13 The global behavior of the Lotka–Volterra system

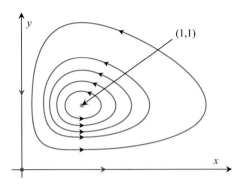

After a simple integration, we find that

$$\frac{x}{e^x} \cdot \frac{y}{e^y} = C, \tag{7.24}$$

where C is a positive constant. Thus, the integral curves of the solution is the graph of the function in (7.24) for different values of C as shown in ◻ Fig. 7.13.

Exercise 7.3

Show that the system

$$\begin{cases} \dfrac{dx}{dt} = y + x^3, \\[2mm] \dfrac{dy}{dt} = x + y + y^3, \end{cases} \tag{7.25}$$

has no limit cycle.

Solution

Let us consider the vector field $V(x, y) = (y + x^3, x + y + y^3)$. Then, we have

$$\mathbf{div}\, V(x, y) = 3x^2 + 3y^2 + 1 \neq 0.$$

Hence, using Bendixson's theorem (Theorem 7.1.2), we deduce that system (7.25) has no limit cycle.

Exercise 7.4

Find the limit cycles of the system

$$\begin{cases} \dfrac{dx}{dt} = \mu x + y + \alpha x(x^2 + y^2), \\[2mm] \dfrac{dy}{dt} = -x + \mu y + \alpha y(x^2 + y^2), \end{cases} \tag{7.26}$$

where μ and α are real parameters.

Solution

We discuss the two cases $\alpha > 0$ and $\alpha < 0$. The second case makes the nonlinear terms as dissipative terms. So, we expect to get a stable limit cycle only for $\alpha < 0$. Without loss of generality, we may discuss the two cases $\alpha = 1$ and $\alpha = -1$.

It is clear that the point $(0, 0)$ is the critical point of (7.26).

Fig. 7.14 The unstable limit cycle for $\mu < 0$ and $\alpha > 0$

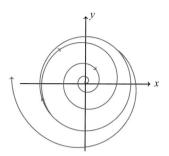

Fig. 7.15 The stable limit cycle for $\mu > 0$ and $\alpha < 0$

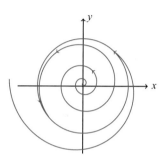

First for $\alpha = 1$, using the polar coordinates, we write

$$x(t) = r \cos \theta, \quad \text{and} \quad y(t) = r \sin \theta,$$

Since

$$x^2 + y^2 = r^2 \quad \text{and} \quad \frac{y}{x} = \tan \theta,$$

then we may easily show that

$$r' = \frac{xx' + yy'}{r} \quad \text{and} \quad \theta' = \frac{xy' - yx'}{r^2}.$$

Then, we may rewrite system (7.26) in the polar coordinates as

$$\begin{cases} r' = \mu r + r^3, \\ \theta' = -1. \end{cases} \tag{7.27}$$

It is clear that for $\mu < 0$ and for $r = \sqrt{-\mu}$, then we have $r' = 0$. This means that the circle of center $\sqrt{-\mu}$ is the only closed orbit. For $r > \sqrt{-\mu}$, then $r' > 0$, which means that the trajectories goes away from the limit cycle. On the other hand for $r < \sqrt{-\mu}$, then the trajectory spirals towards the origin. Since $\theta' = -1$, then the direction of the trajectories are clockwise. See ◘ Fig. 7.14.

Second for $\alpha = -1$, then system (7.27) becomes

$$\begin{cases} r' = \mu r - r^3, \\ \theta' = -1. \end{cases} \tag{7.28}$$

Hence, for $\mu > 0$, the trajectories go to the limit cycle $K : x^2 + y^2 = \mu$ in both cases for $r < \sqrt{\mu}$ and for $\mu > \sqrt{\mu}$. This means that the limit cycle K is a stable limit cycle as in ◘ Fig. 7.15.

Servicepart

References

1. W.E. Boyce, R.C. DiPrima, *Elementary Differential Equations and Boundary Value Problems* (Wiley, 2008)

2. D.N. Burghes, M.S. Borrie, *Modelling with Differential Equations* (E. Horwood, 1981)

3. H. Cartan, *Calcul différentiel* (Hermann, Paris, 1967)

4. S.D. Chatterji, *Cours d'analyse, tome 3. Équations différentielles ordinaires et aux dérivées partielles* [Ordinary and Partial Differential Equations] (Presses Polytechniques et Universitaires Romandes, Lausanne, 1998)

5. L. Debnath, D. Bhatta, *Integral Transforms and Their Applications*, 2nd edn. (Chapman & Hall/CRC, Boca Raton, FL, 2007)

6. J.-P. Demailly, *Analyse numérique et équations différentielles, novelle édition* (EDP sciences, 2012)

7. C.H. Edwards, D.E. Penney, *Instructor's Solutions Manual, Elementary Differential Equations*, 6th edn. (Prentice Hall, 2000)

8. J.K. Hale, *Introduction to Functional Differential Equations*. Applied Mathematical Sciences, vol. 99 (Springer, 1993)

9. N.J. Higham, *Functions of Matrices: Theory and Computation* (Siam, 2008)

10. M.W. Hirsch, S. Smale, R.L. Devaney, *Differential Equations, Dynamical Systems, and an Introduction to Chaos*, 2nd edition. Pure and Applied Mathematics (Amsterdam), vol. 60 (Elsevier/Academic Press, Amsterdam, 2004)

11. H.K. Khalil, *Nonlinear Systems*, (Macmillan Publishing Company, New York, 1992)

12. J.M. Knudsen, P.G. Hjorth, *Elements of Newtonian Mechanics: Including Nonlinear Dynamics* (Springer, 2000)

13. J.F.P. Martin, Some results on matrices which commute with their derivatives, SIAM Journal on Applied Mathematics **15**(5), 1171–1183 (1967)

14. A. Mattuck, Differential equations. (MIT Open Courseware, 2006), http://ocw.mit.edu/18-03S06. Accessed January, 2015

15. R.K. Nagle, E.B. Saff, A.D. Snider, B. West, *Fundamentals of Differential Equations and Boundary Value Problems*, 6th edn. (Addison-Wesley New York, 2012)

16. S.A. Orszag, C.M. Bender, *Advanced Mathematical Methods for Scientists and Engineers* (McGraw-Hill, 1978)

17. L. Perko, *Differential Equations and Dynamical Systems*, 3rd edn. Texts in Applied Mathematics, vol. 7 (Springer, New York, 2001)

18. E.J. Putzer, Avoiding the Jordan canonical form in the discussion of linear systems with constant coefficients, Amer. Math. Monthly **73**, 2–7 (1966)

19. J. Quinet, *Cours élémentaire de mathématiques supérieures, tome 4. Équations différentielles*, 6th edn. (Dunod, 1977)

20. D.A. Sánchez, *Ordinary Differential Equations and Stability Theory. An Introduction* (Dover Publications, New York, 1979). Reprint of the 1968 original

21. B.E. Shapiro, *Lecture Notes in Differential equations* (Lulu.com, 2011)

22. G.F. Simmons, *Differential Equations with Applications and Historical Notes* (McGraw-Hill, 1972)

23. S. Smale, Mathematical problems for the next century, The Mathematical Intelligencer, **20**(2), 7–15 (1998)

24. G. Teschl, *Ordinary Differential Equations and Dynamical Systems*. Graduate Studies in Mathematics, vol. 140 (American Mathematical Soc., 2012)

25. M.D. Weir, G.B. Hass, J. Thomas, *Thomas' Calculus*, 11th edn. (Addison-Wesley, 2010)

26. Z. Zhang, On the uniqueness of limit cycles of certain equations of nonlinear oscillations, Dokl. Akad. Nauk SSSR **119**, 659–662 (1958)

Index